Editor
Gisela Lee, M.A.

Managing Editor
Karen Goldfluss, M.S. Ed.

Editor-in-Chief
Sharon Coan, M.S. Ed.

Cover Artist
Barb Lorseyedi

Art Coordinator
Kevin Barnes

Art Director
CJae Froshay

Imaging
Alfred Lau
James Edward Grace
Rosa C. See

Product Manager
Phil Garcia

Publisher
Mary D. Smith, M.S. Ed.

Author

Robert Smith

Teacher Created Resources, Inc.
6421 Industry Way
Westminster, CA 92683
www.teachercreated.com

ISBN-0-7439-8625-3

Reprinted, 2005
Made in U.S.A.

Table of Contents

Introduction

The old adage "practice makes perfect" can really hold true for your child and his or her education. The more practice and exposure your child has with concepts being taught in school, the more success he or she is likely to find. For many parents, knowing how to help your children can be frustrating because the resources may not be readily available. As a parent it is also difficult to know where to focus your efforts so that the extra practice your child receives at home supports what he or she is learning in school.

This book has been designed to help parents and teachers reinforce basic skills with their children. *Practice Makes Perfect* reviews basic math skills for children in grade 5. The math focus is geometry. While it would be impossible to include all concepts taught in grade 5 in this book, the following basic objectives are reinforced through practice exercises. These objectives support math standards established on a district, state, or national level. (Refer to the Table of Contents for the specific objectives of each practice page.)

- describing and classifying angles
- identifying congruent and similar figures
- drawing lines of symmetry
- identifying flips, turns, and slides
- describing triangles
- measuring angles in a triangle
- naming plane and solid geometric figures
- identifying faces, edges, and vertices
- finding perimeter and area of polygons, triangles, circles, etc.
- finding the radius and diameter of a circle

There are 36 practice pages organized sequentially, so children can build their knowledge from more basic skills to higher-level math skills. (**Note:** Have children show all work where computation is necessary to solve a problem. For multiple choice responses on practice pages, children can fill in the letter choice or circle the answer.) Also note that not all shapes are drawn to scale. Following the practice pages are six practice tests. These provide children with multiple-choice test items to help prepare them for standardized tests administered in schools. As your child completes each test, he or she should fill in the correct bubbles on the answer sheet (page 46). To correct the test pages and the practice pages in this book, use the answer key provided on pages 47 and 48.

How to Make the Most of This Book

Here are some useful ideas for optimizing the practice pages in this book:

- Set aside a specific place in your home to work on the practice pages. Keep it neat and tidy with materials on hand.
- Set up a certain time of day to work on the practice pages. This will establish consistency. An alternative is to look for times in your day or week that are less hectic and conducive to practicing skills.
- Keep all practice sessions with your child positive and constructive. If the mood becomes tense, or you and your child are frustrated, set the book aside and look for another time to practice with your child.
- Help with instructions if necessary. If your child is having difficulty understanding what to do or how to get started, work through the first problem through with him or her.
- Review the work your child has done. This serves as reinforcement and provides further practice.
- Allow your child to use whatever writing instruments he or she prefers. For example, colored pencils can add variety and pleasure to drill work.
- Pay attention to the areas in which your child has the most difficulty. Provide extra guidance and exercises in those areas. Allowing children to use drawings and manipulatives, such as coins, tiles, game markers, or flash cards, can help them grasp difficult concepts more easily.
- Look for ways to make real-life applications to the skills being reinforced.

Practice 1

Reminders

- An acute angle measures less than 90°.
- An obtuse angle measures more than 90° and less than 180°.
- A right angle measures exactly 90°.
- A straight angle measures exactly 180°.

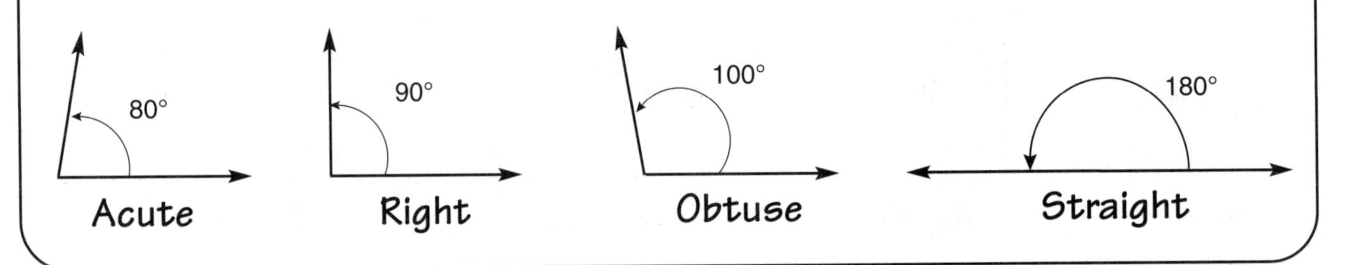

Directions: Label each of these angles as acute, right, obtuse, or straight angles.

1. 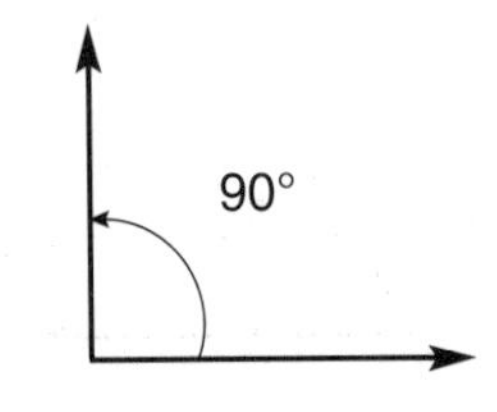

____

2. 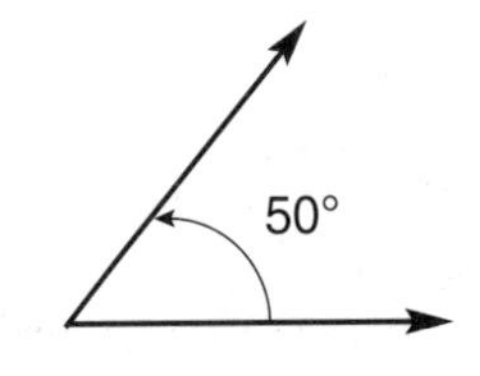

____

3. 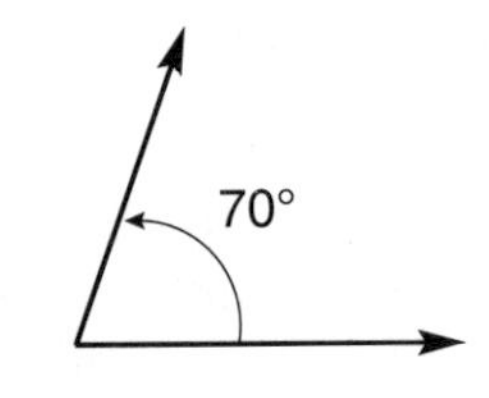

____

4. 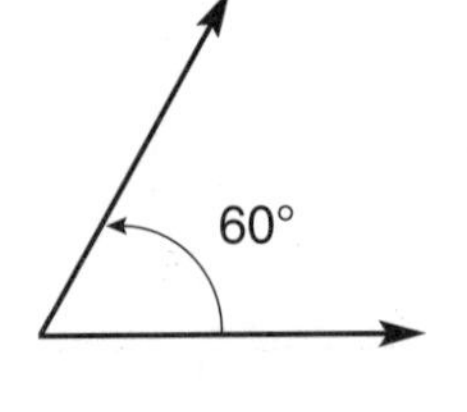

____

5. 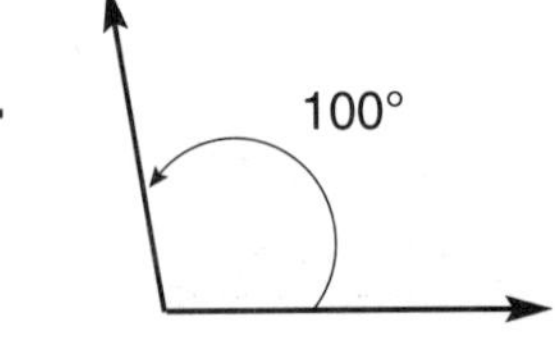

____

6. 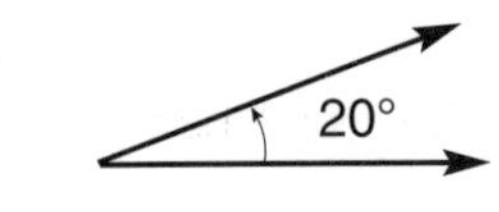

____

7. 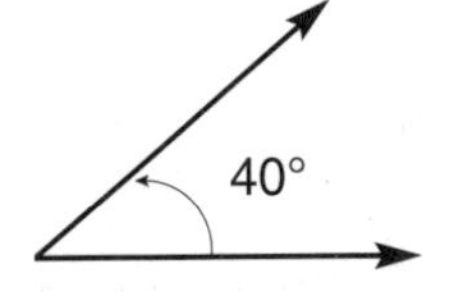

____

8. 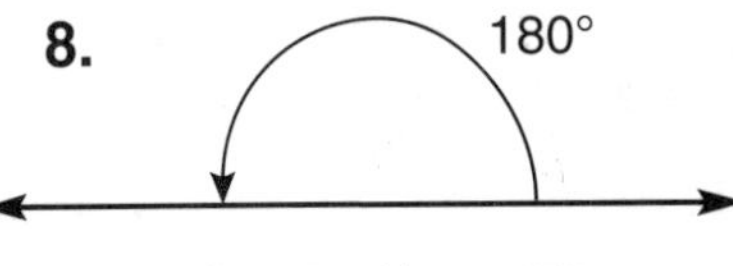

____

9. 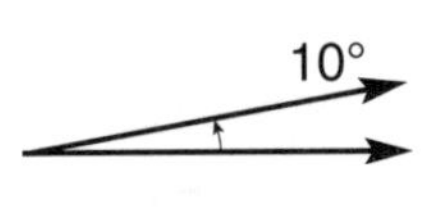

____

10. 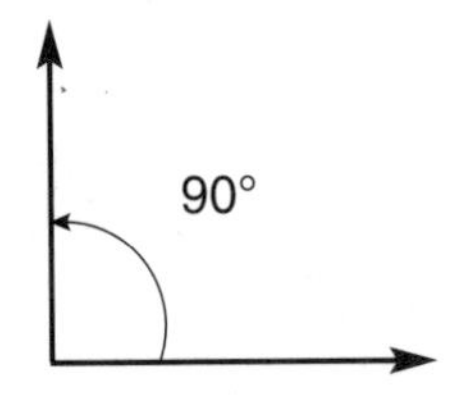

____

11. 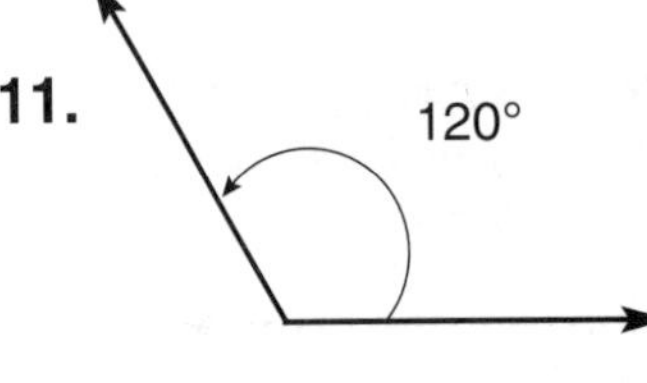

____

12. 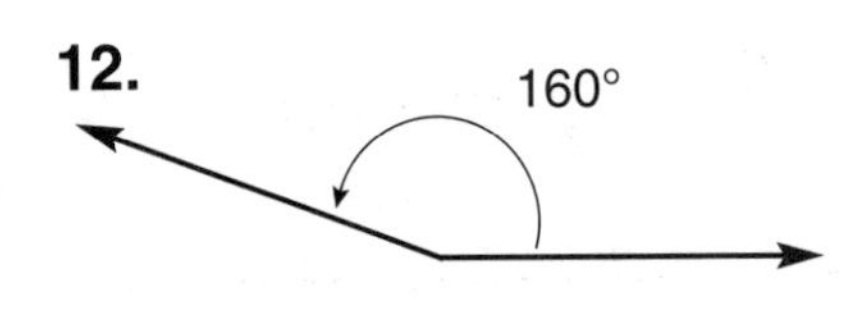

____

Practice 2

Reminders

- An acute angle measures less than 90°.
- An obtuse angle measures more than 90° and less than 180°.
- A right angle measures exactly 90°.
- A straight angle measures exactly 180°.

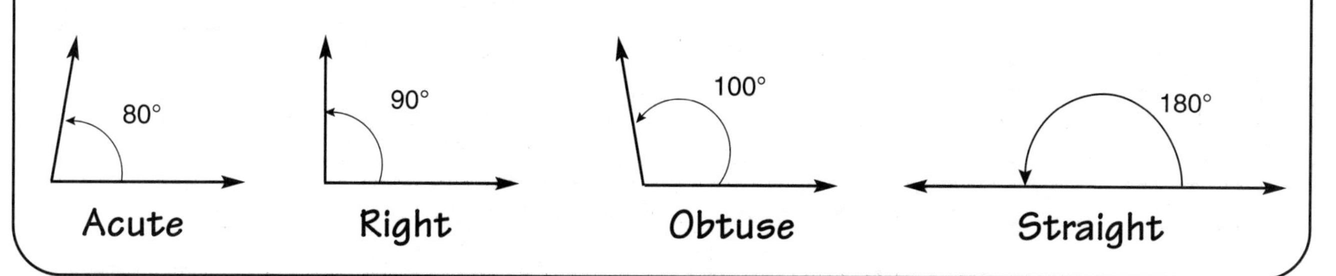

Directions: Label each of these angles as acute, right, obtuse, or straight angles.

1.

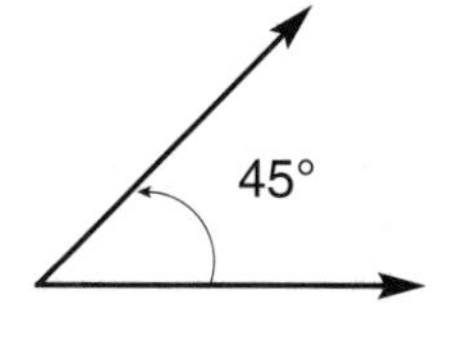

____

2.

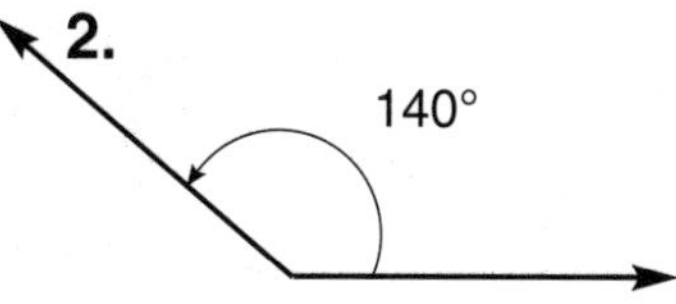

____

3.

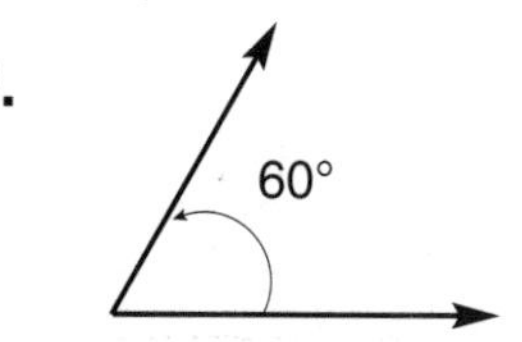

____

4.

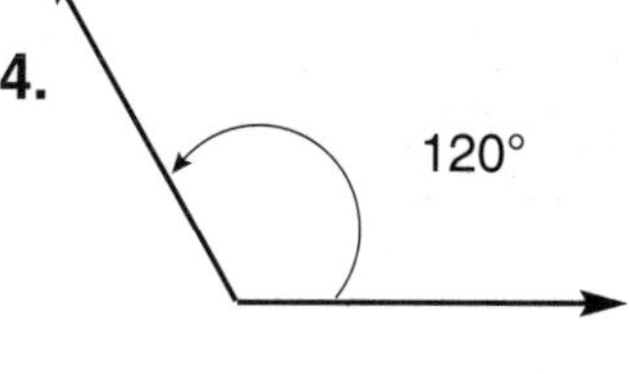

____

5.

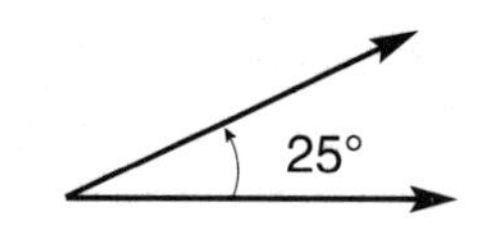

____

6.

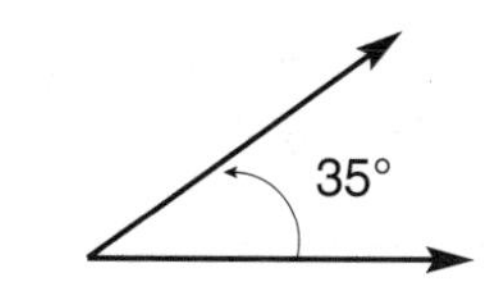

____

7.

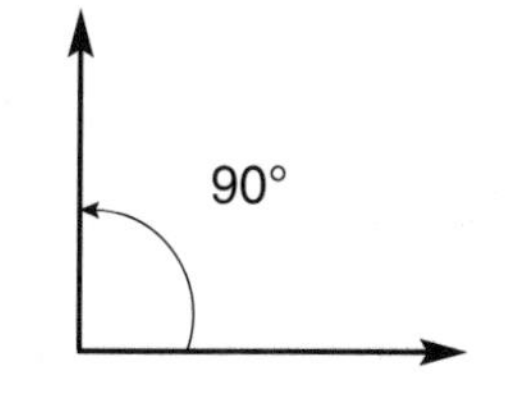

____

8.

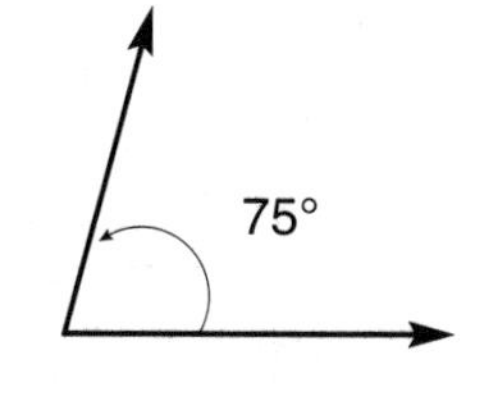

____

9.

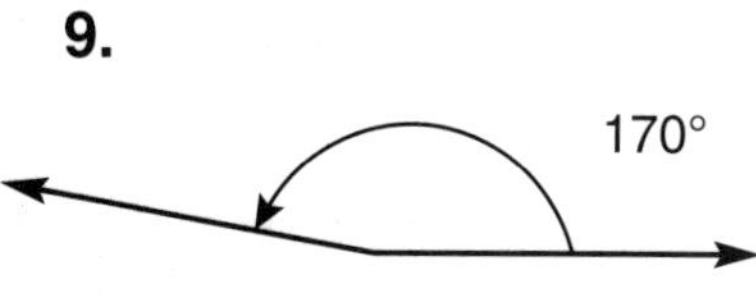

____

10.

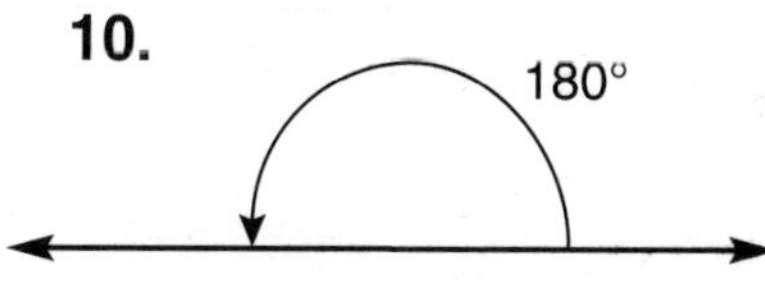

____

11.

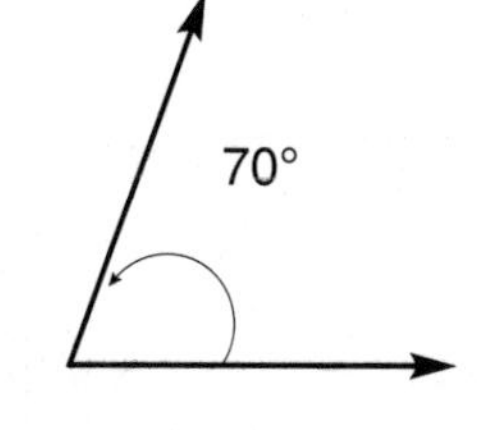

____

12.

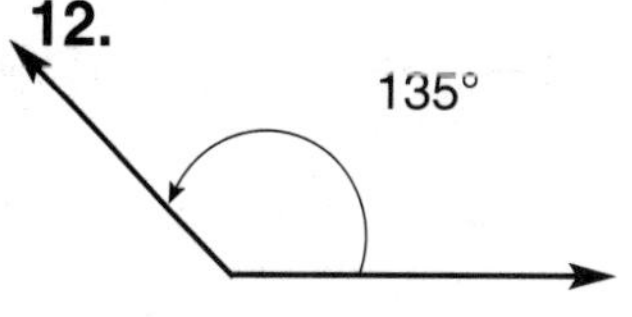

____

Practice 3

Directions: Use a protractor to measure each of the angles below. Write the number of degrees and the name of each angle: acute, right, obtuse, or straight.

1.

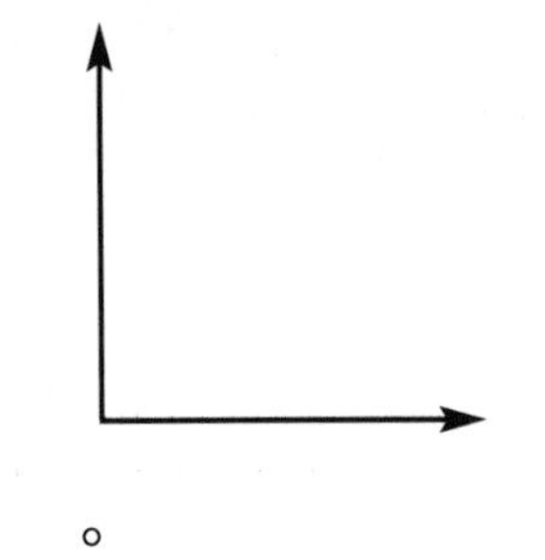

____° ____________

2.

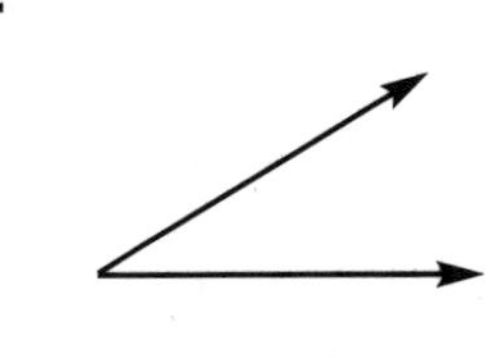

____° ____________

3.

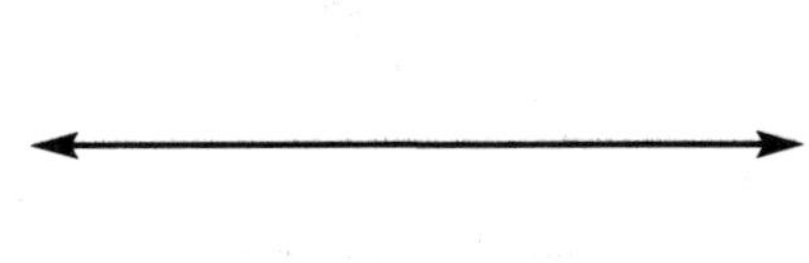

____° ____________

4.

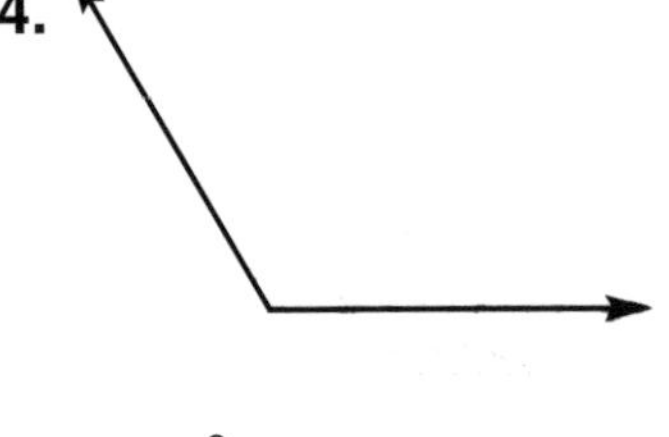

____° ____________

5.

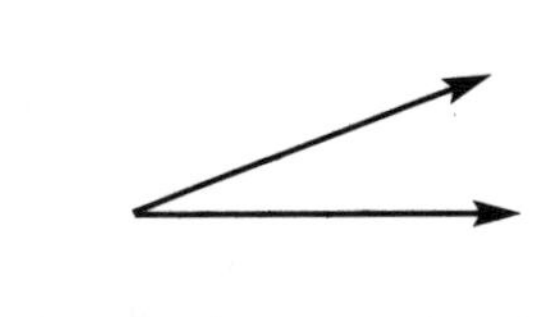

____° ____________

6.

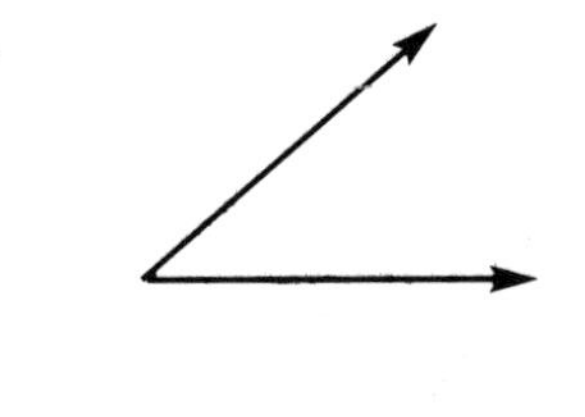

____° ____________

7.

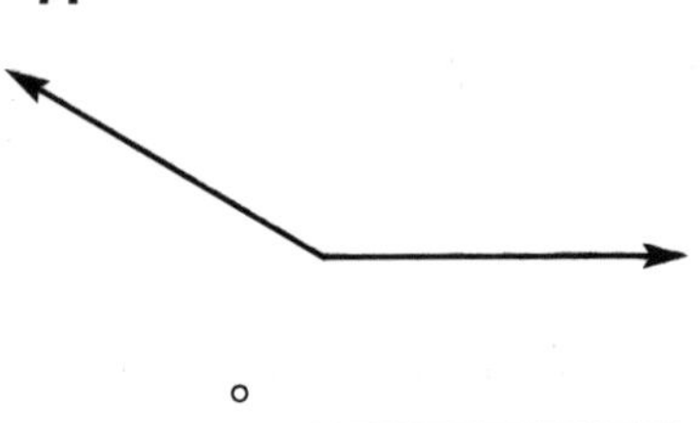

____° ____________

8.

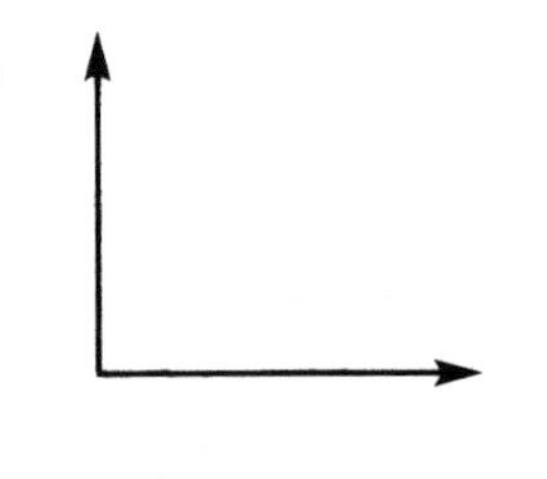

____° ____________

9.

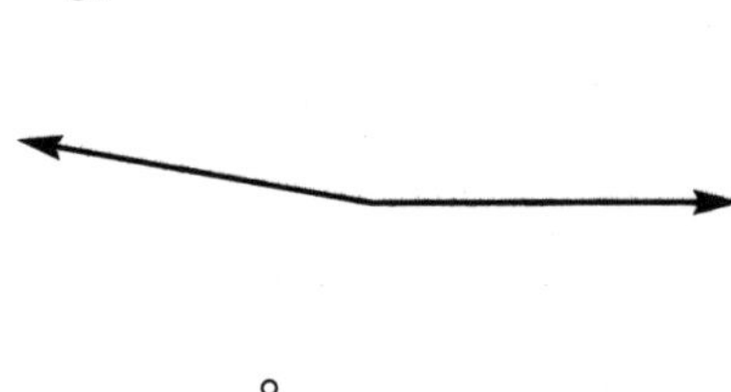

____° ____________

10.

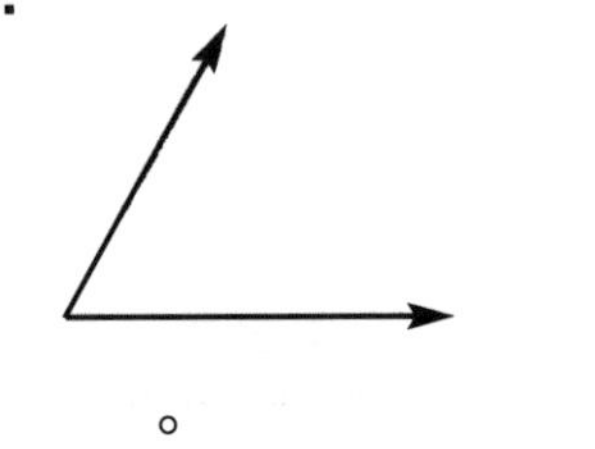

____° ____________

11.

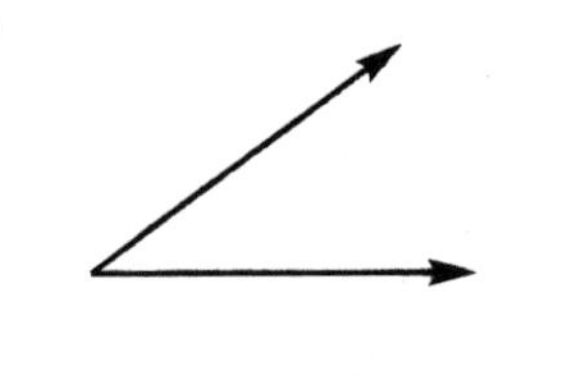

____° ____________

12.

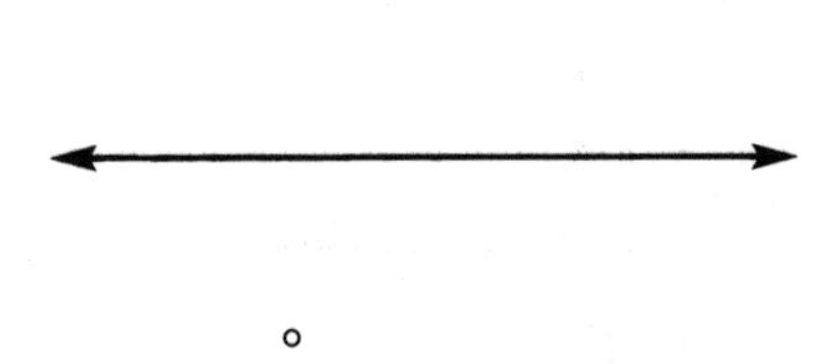

____° ____________

Practice 4

Directions: Use a protractor to measure each of the angles below. Write the number of degrees and the name of each angle: acute, right, obtuse, or straight.

1.

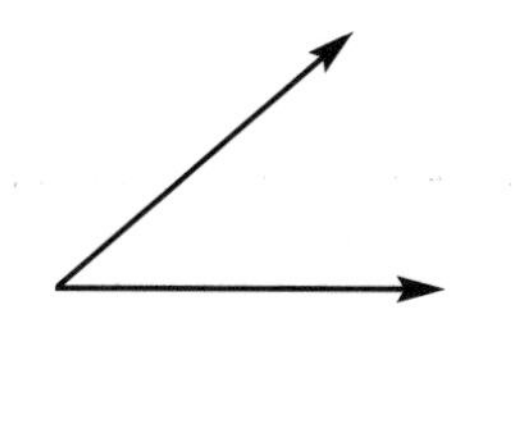

_____° ______________

2.

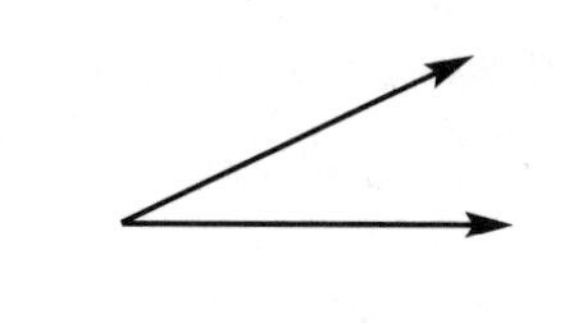

_____° ______________

3.

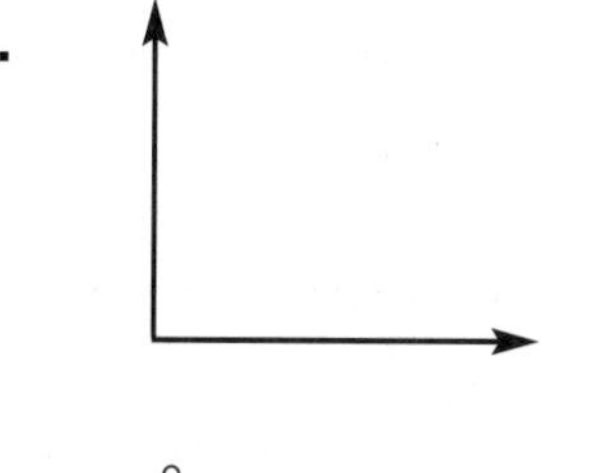

_____° ______________

4.

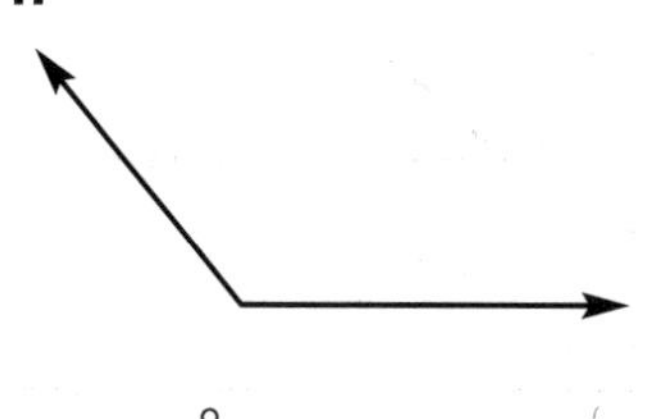

_____° ______________

5.

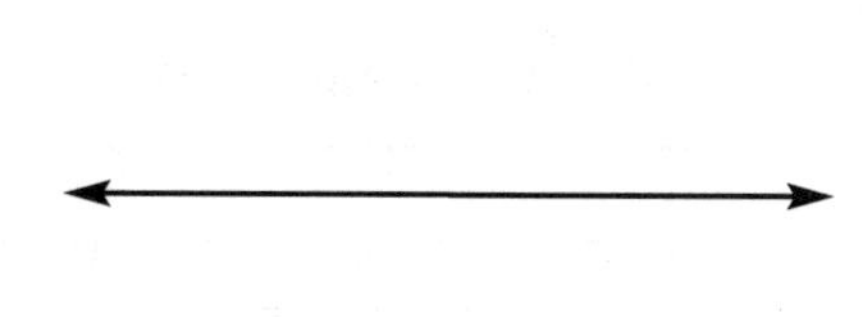

_____° ______________

6.

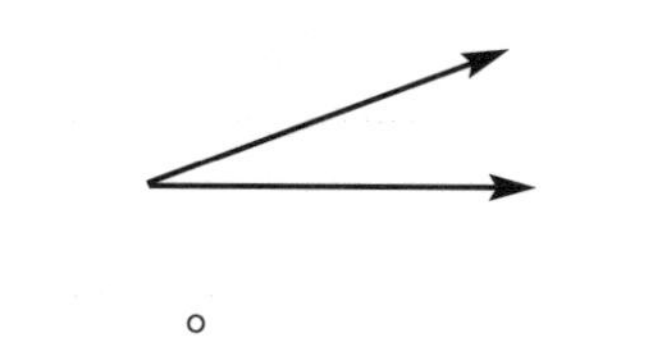

_____° ______________

7.

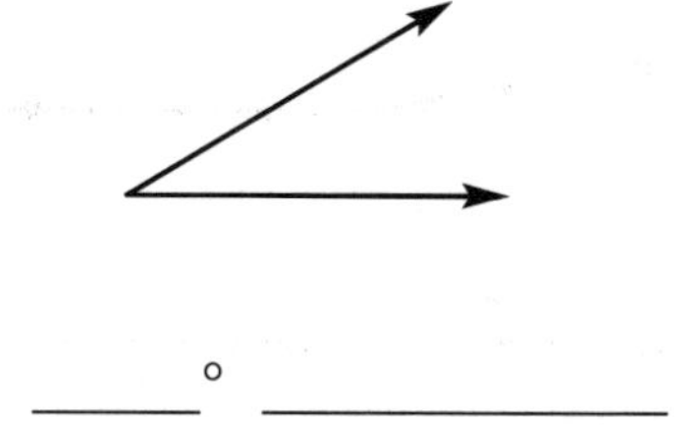

_____° ______________

8.

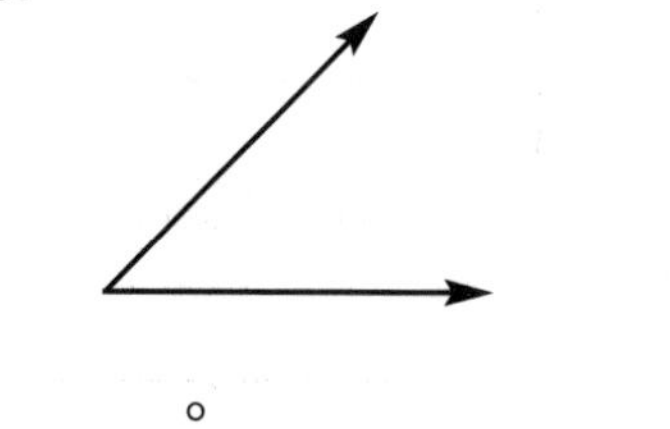

_____° ______________

9.

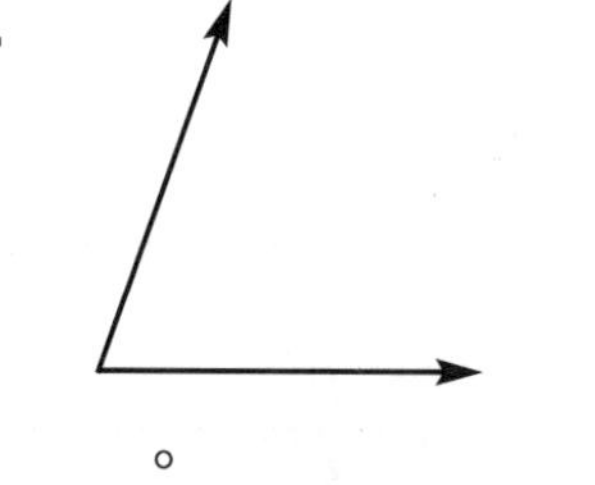

_____° ______________

10.

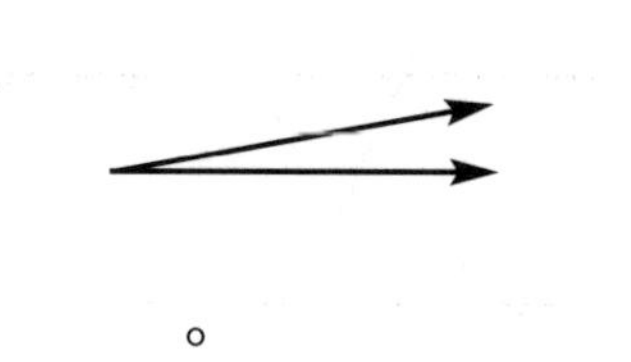

_____° ______________

11.

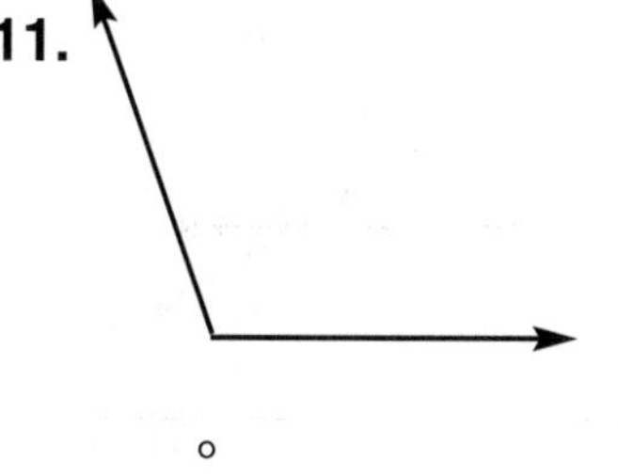

_____° ______________

12.

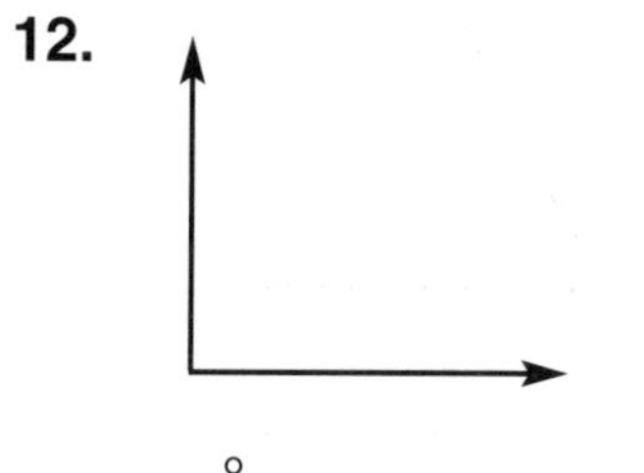

_____° ______________

Practice 5

        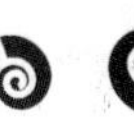 

Reminders

- A right triangle has one 90° angle.
- An equilateral triangle has three equal sides and three equal angles of 60° each.
- An isosceles triangle has two equal sides and two equal angles.
- A scalene triangle has no equal sides and no equal angles.
- An isosceles right triangle has one 90° angle and two 45° angles. The sides adjacent (next to) the right angle are equal.
- An acute triangle has all three angles less than 90°.
- An obtuse triangle has one angle greater than 90°.
- Triangles can have more than one name.

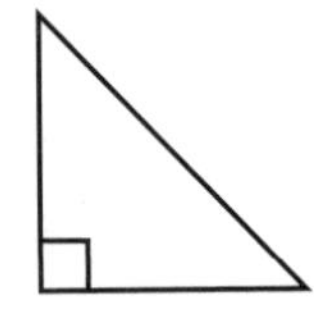 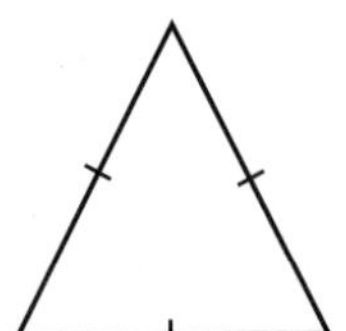 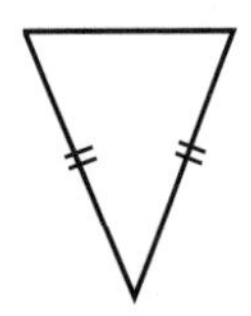 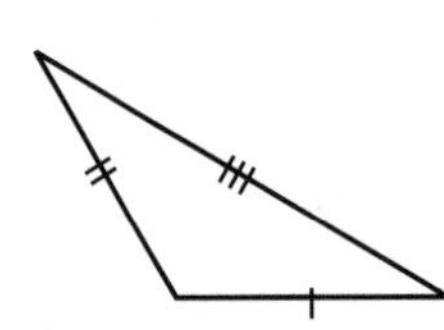 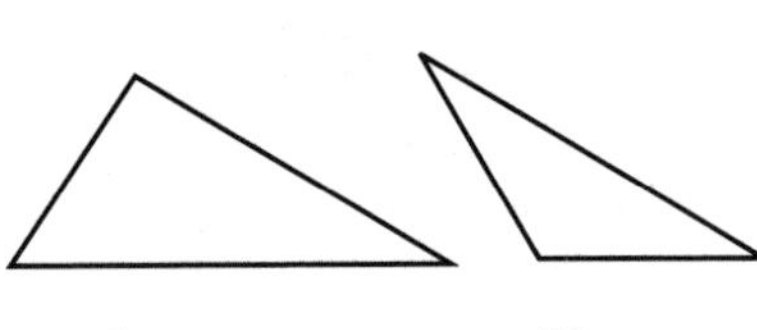

Right | Equilateral | Isosceles | Scalene | Acute | Obtuse

Directions: Identify each triangle. If the triangle has more than one name, use both names.

1. 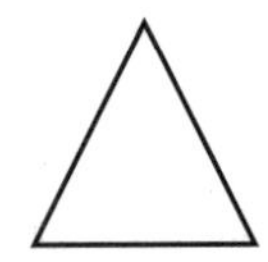

____

2. 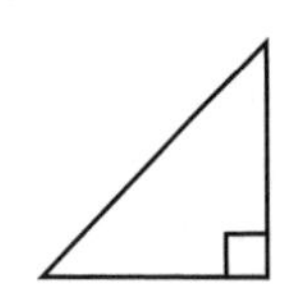

____

3. 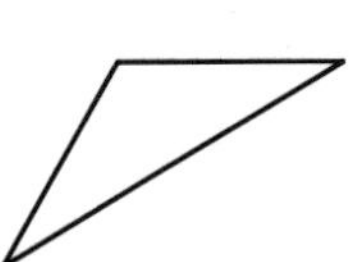

____

4. 

____

5. 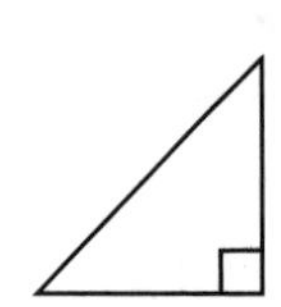

____

6. 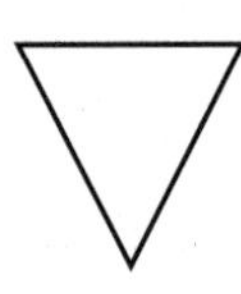

____

7. 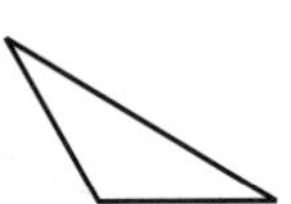

____

8. 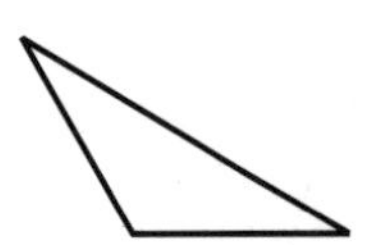

____

9. 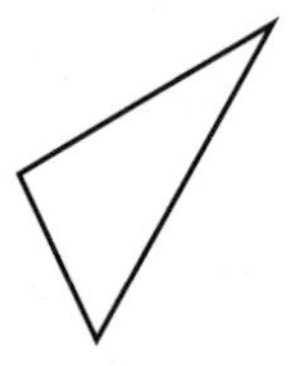

____

Practice 6

Reminders

- The sum of the interior angles of every triangle is 180°.
- If you know two of the angles of a triangle, you can find the third angle by adding the two angles you know and subtracting the sum from 180°.

Directions: Compute the number of degrees in each unmarked angle.

1.

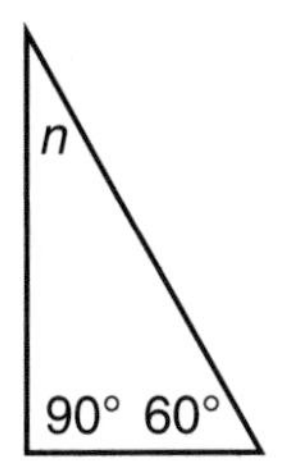

$n°$ = ____________________

2.

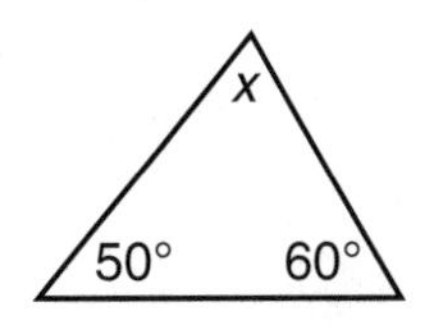

$x°$ = ____________________

3.

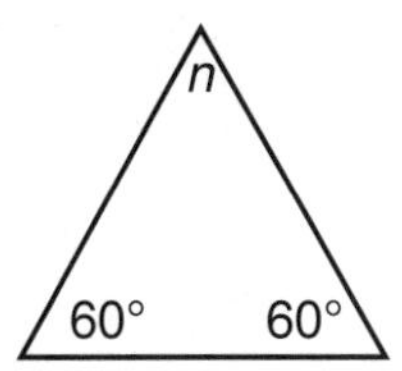

$n°$ = ____________________

4.

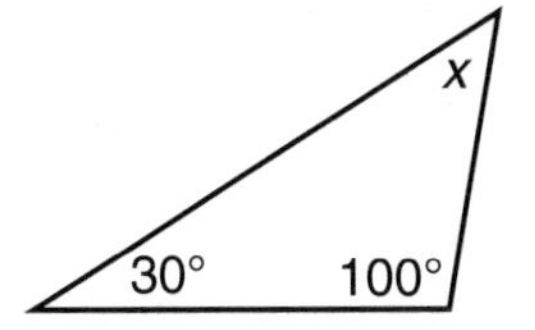

$x°$ = ____________________

5.

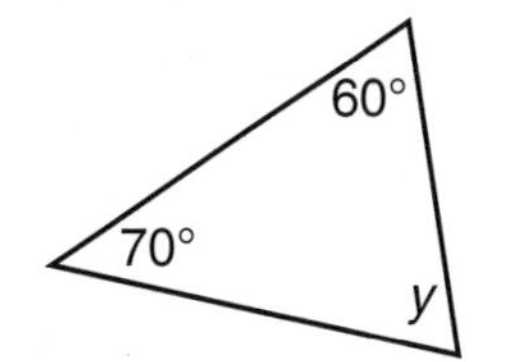

$y°$ = ____________________

6.

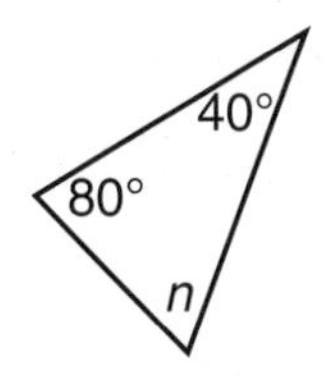

$n°$ = ____________________

7.

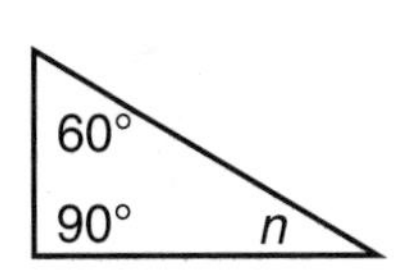

$n°$ = ____________________

8.

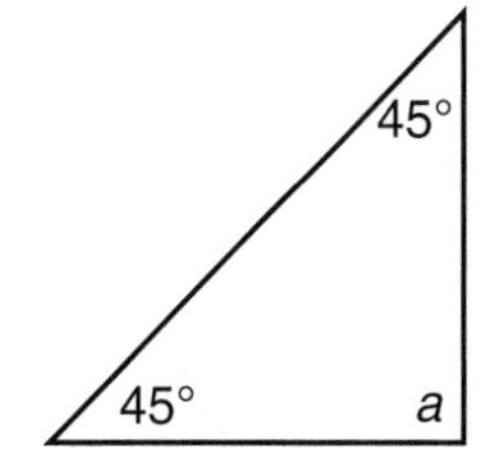

$a°$ = ____________________

9.

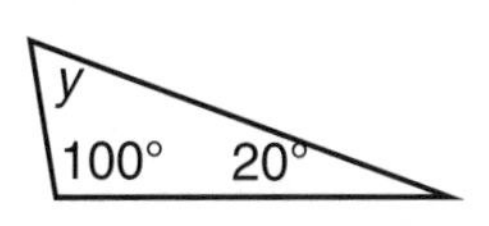

$y°$ = ____________________

10.

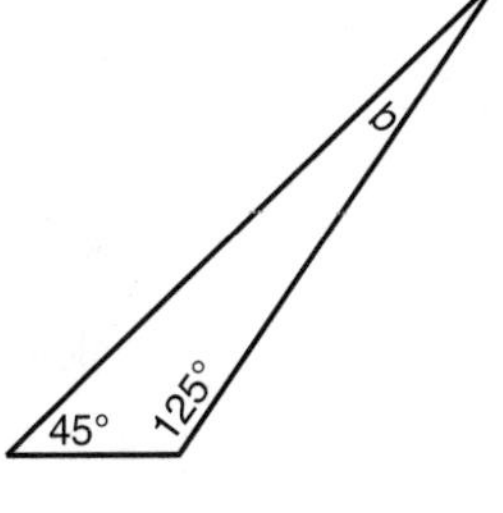

$b°$ = ____________________

11.

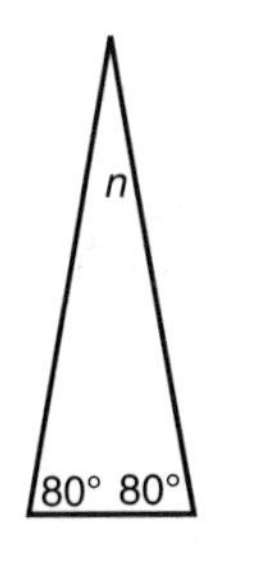

n = ____________________

12.

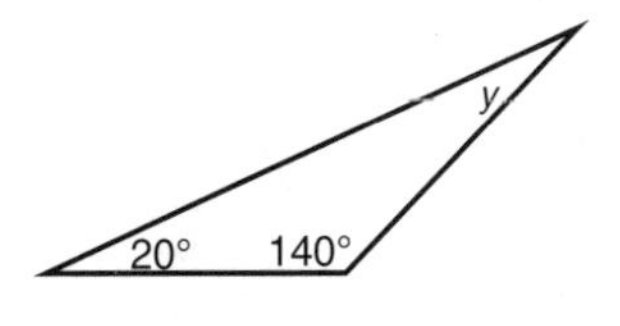

$y°$ = ____________________

Practice 7

Reminders

- The sum of the interior angles of every triangle is 180°.
- If you know two of the angles of a triangle, you can find the third angle by adding the two angles you know and subtracting the sum from 180°.

Directions: Compute the number of degrees in each unmarked angle.

1.

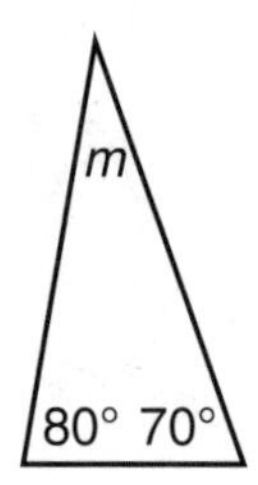

$m° =$ ____________________

2.

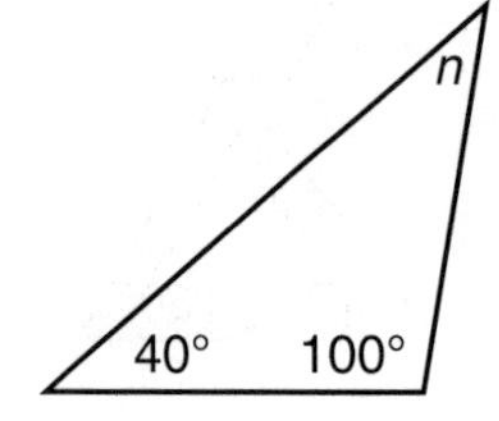

$a° =$ ____________________

3.

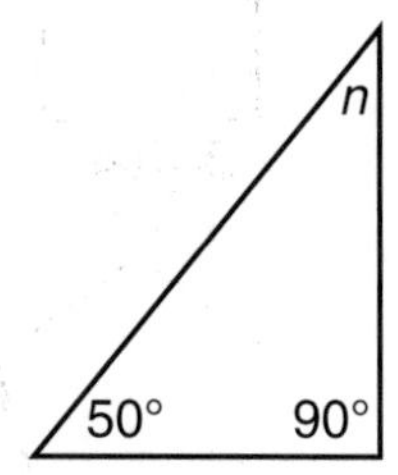

$n° =$ ____________________

4.

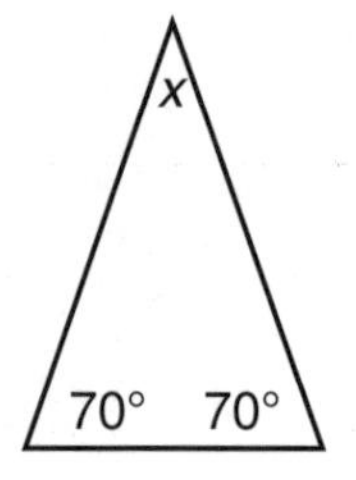

$x° =$ ____________________

5.

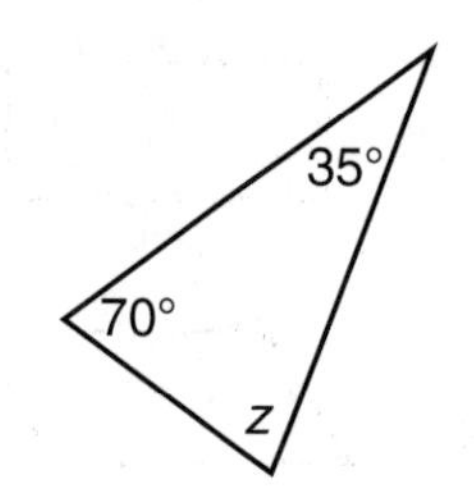

$z° =$ ____________________

6.

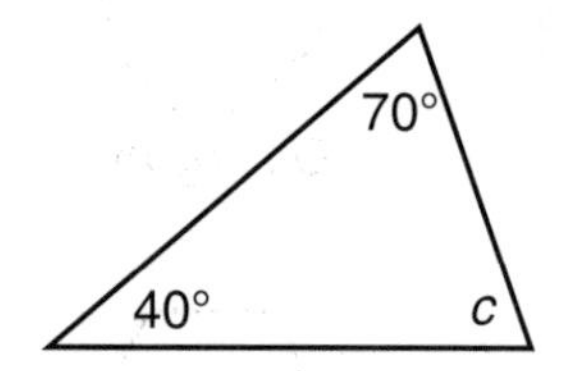

$c° =$ ____________________

7.

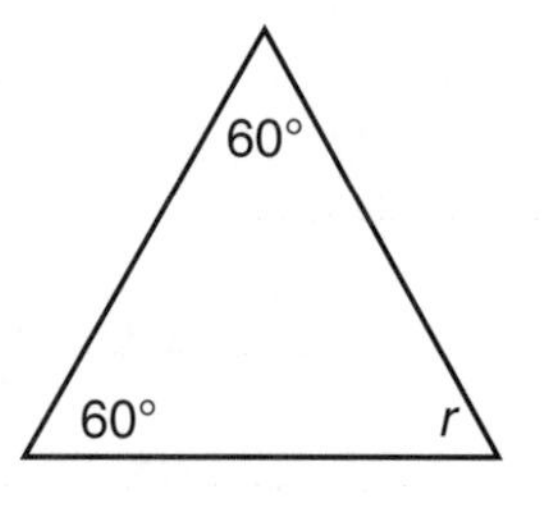

$r° =$ ____________________

8.

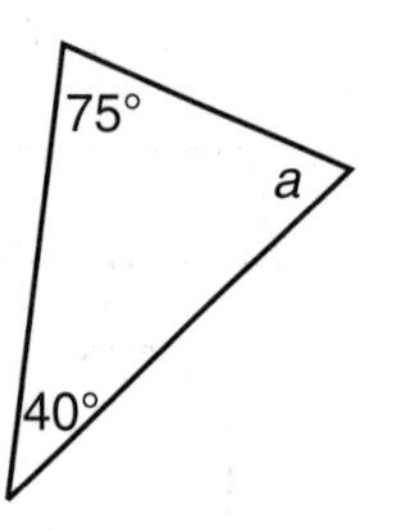

$a° =$ ____________________

9.

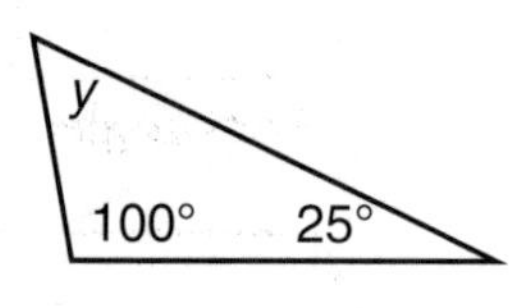

$y° =$ ____________________

10.

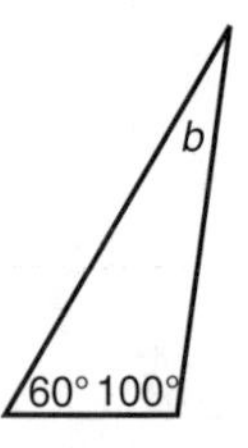

$b° =$ ____________________

11.

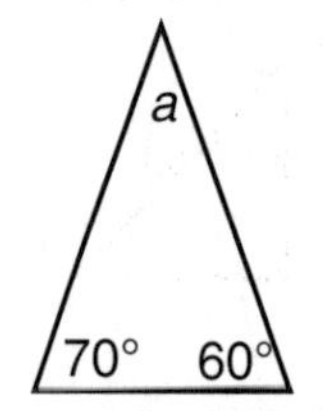

$a =$ ____________________

12.

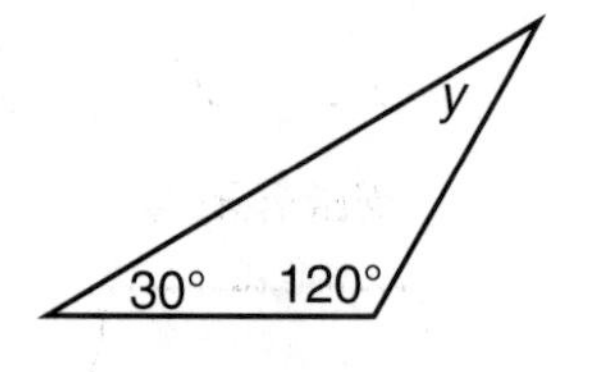

$y° =$ ____________________

Practice 8

Polygon Names

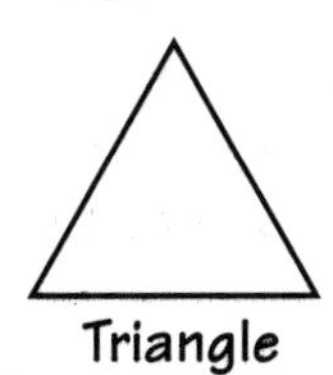

Triangle

Square

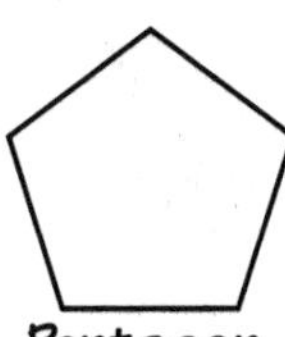

Pentagon

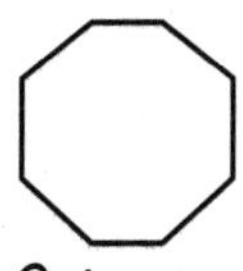

Octagon

Hexagon

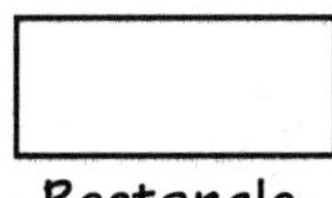

Rectangle

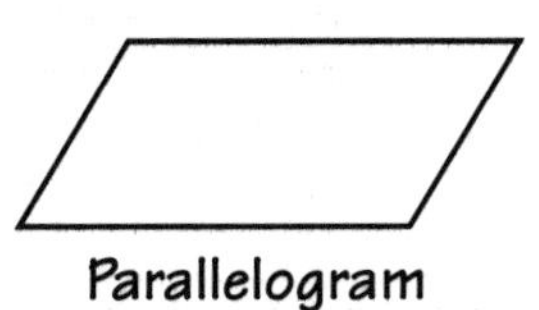

Parallelogram

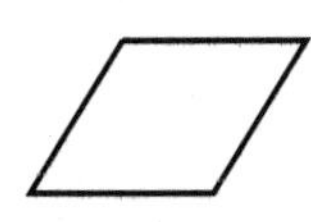

Rhombus

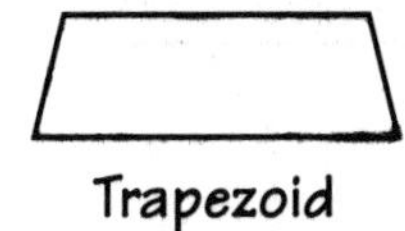

Trapezoid

Directions: Use the names listed on the left to identify each of the polygons below. Use the most specific name for each figure.

1. 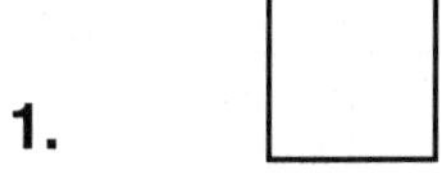 ______________________

2. 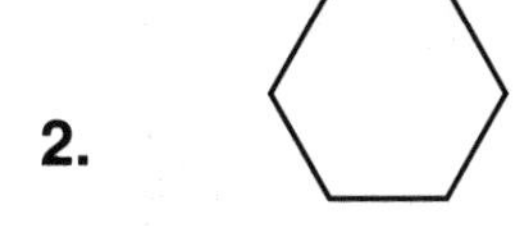 ______________________

3. ______________________

4.  ______________________

5. ______________________

6. 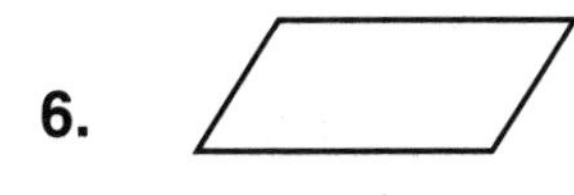 ______________________

7. ______________________

8. 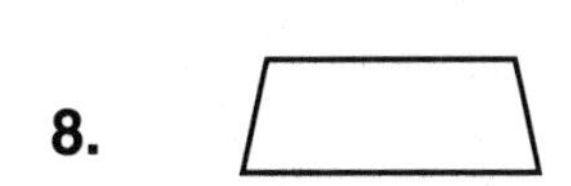 ______________________

9. ______________________

10. 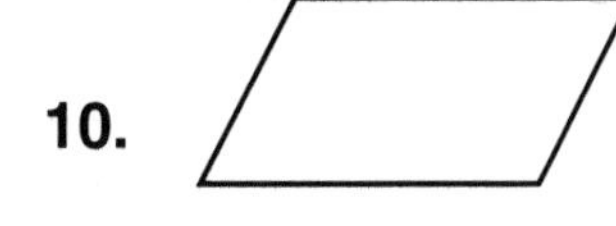 ______________________

11. ______________________

12. ______________________

Practice 9

Reminder

The interior angles of a quadrilateral always add up to 360°.

Directions: Compute the number of degrees in each unmarked angle.

1.

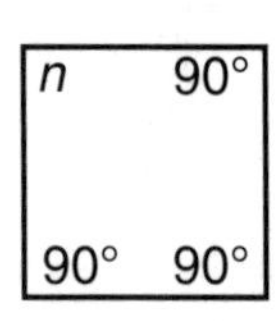

$n° =$ ________________

2.

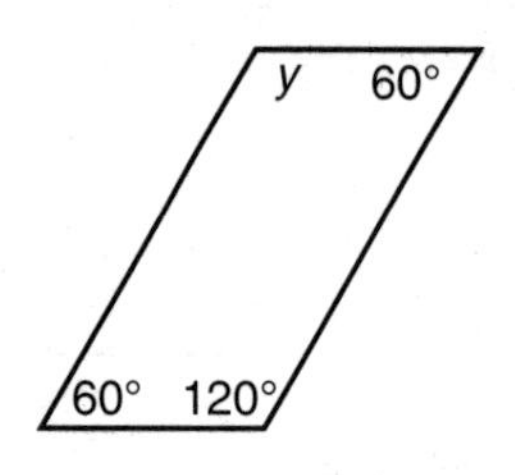

$y° =$ ________________

3.

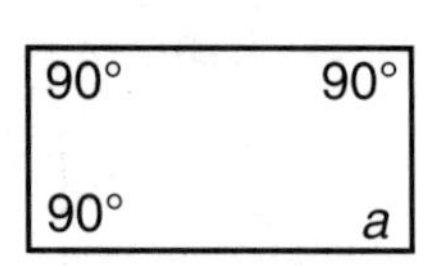

$a° =$ ________________

4.

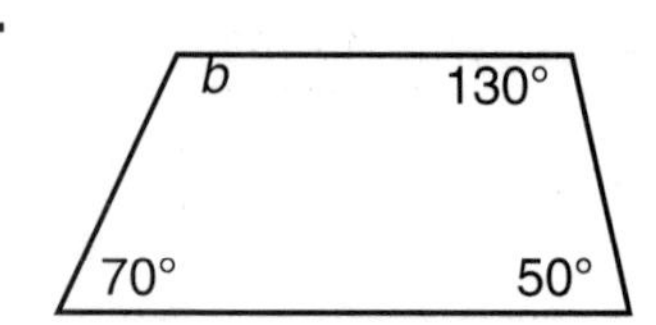

$b° =$ ________________

5.

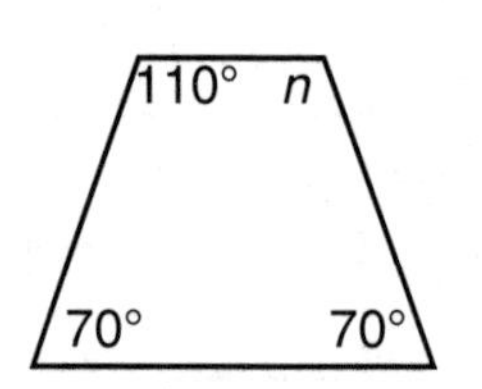

$n° =$ ________________

6.

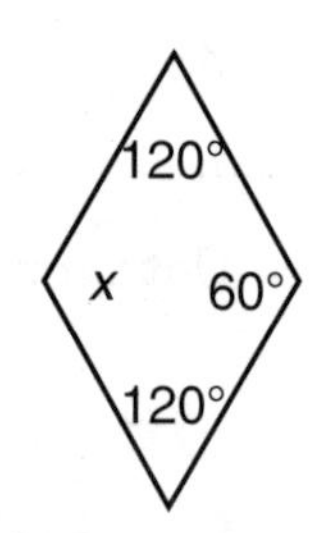

$x° =$ ________________

7.

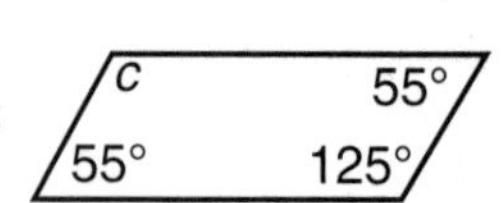

$c° =$ ________________

8.

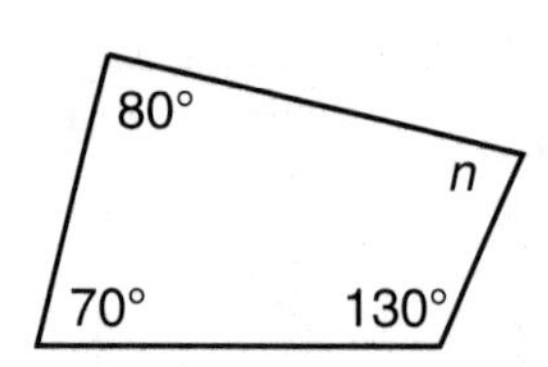

$n° =$ ________________

9.

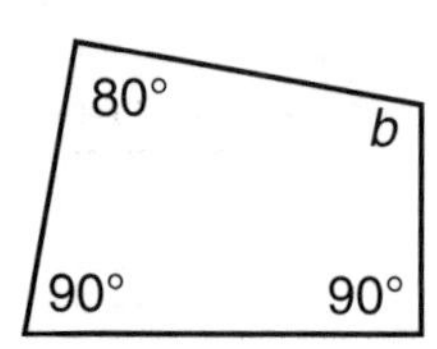

$b° =$ ________________

10.

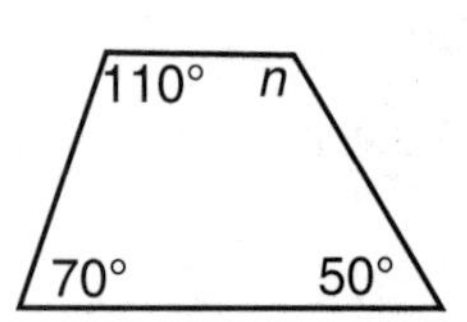

$n° =$ ________________

11.

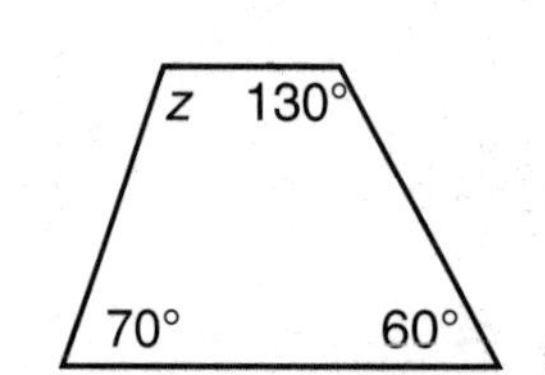

$z° =$ ________________

12.

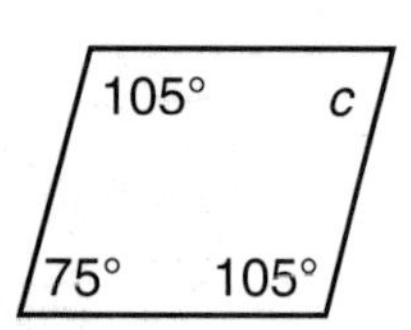

$c° =$ ________________

Practice 10

Reminders

- The perimeter of a geometric figure is the distance around the figure.
- The perimeter of a square can be computed by multiplying the length of one side of the square by four.

Directions: Compute the perimeter of these squares.

1. 8 ft.

____

2. 12 cm

____

3. 10 m

____

4. 7 in.

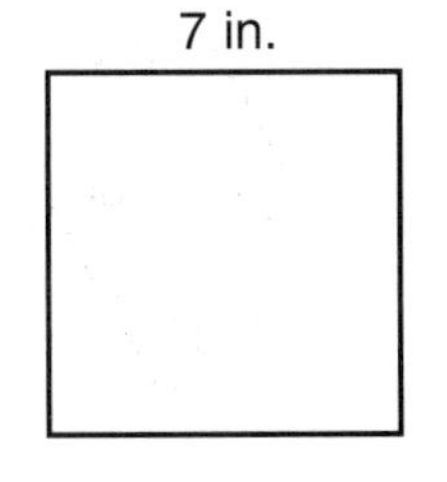

____

5. 12 ft.

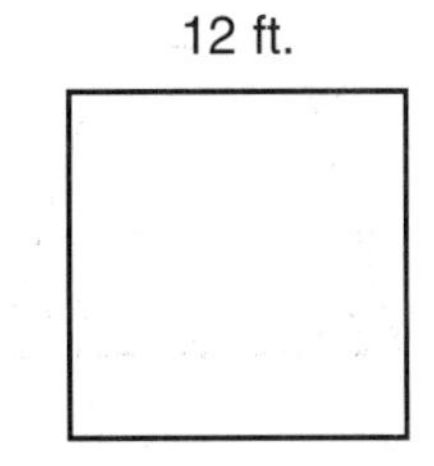

____

6. 13 yards

____

7. 11 cm

____

8. 21 miles

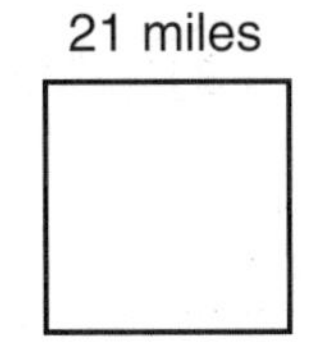

____

9. 44 m

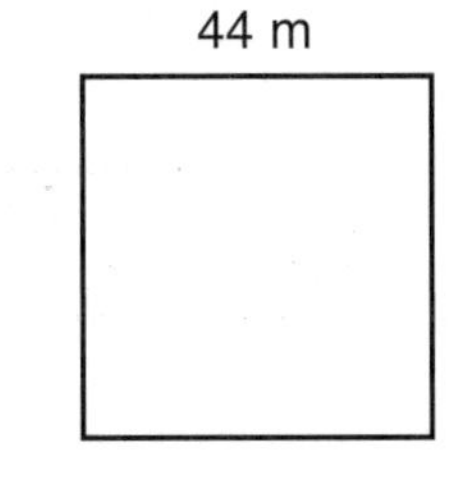

____

10. 77 mm

____

11. 141 ft.

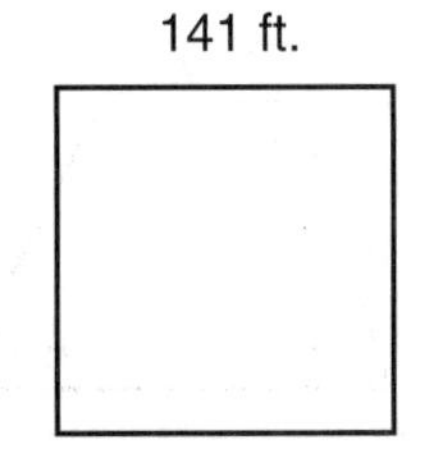

____

12. 368 m

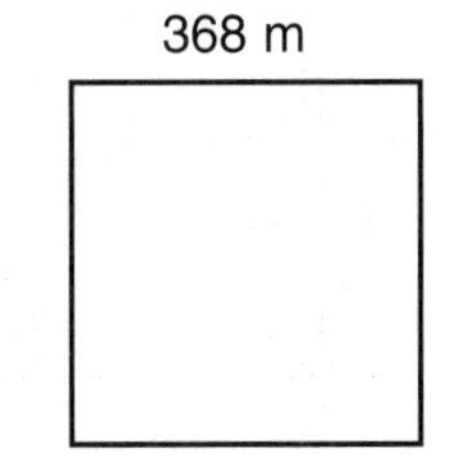

____

Practice 11

Reminder

The perimeter of a rectangle is computed by adding the length plus the width and multiplying the sum times two.

P = 2 x (l + w)

Directions: Compute the perimeter of these rectangles.

1.

12 m

6 m

____

2.

12 ft.

9 ft.

____

3.

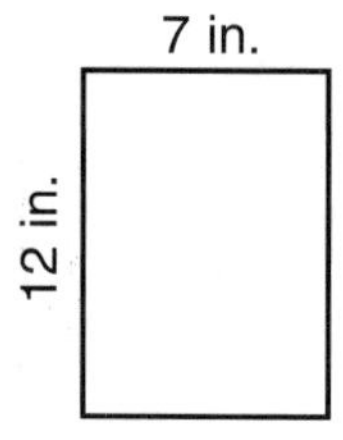

____

4.

20 yards

15 yards

____

5.

18 ft.

10 ft.

____

6.

____

7.

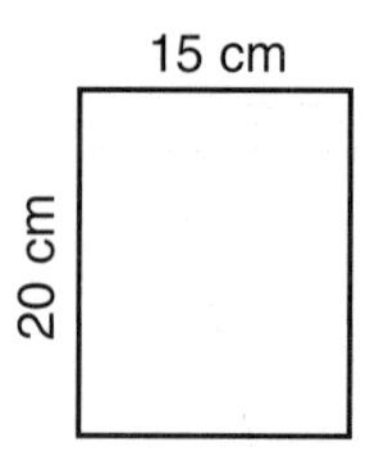

____

8.

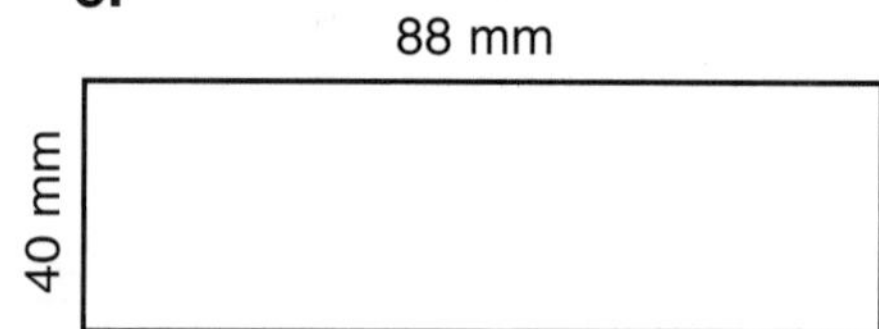

____

9.

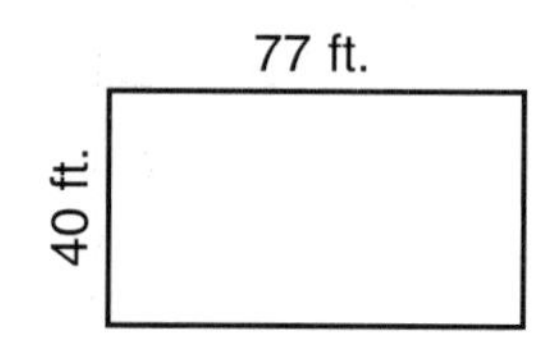

____

10.

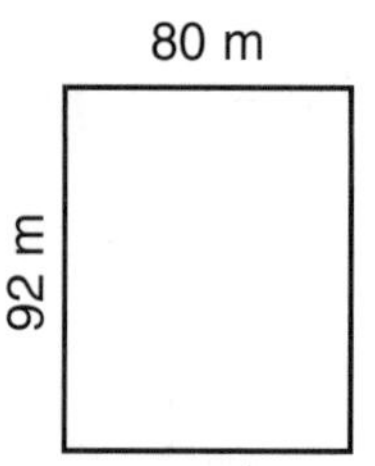

____

11.

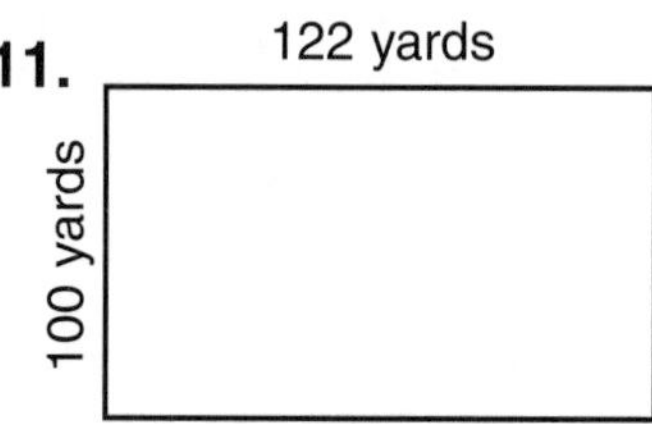

____

12.

144 in.

60 in.

____

Practice 12

Reminder

The perimeter of a parallelogram is computed by adding the length plus the width and multiplying the sum times two.

$$P = 2 \times (l + w)$$

Directions: Compute the perimeter of these parallelograms.

1.

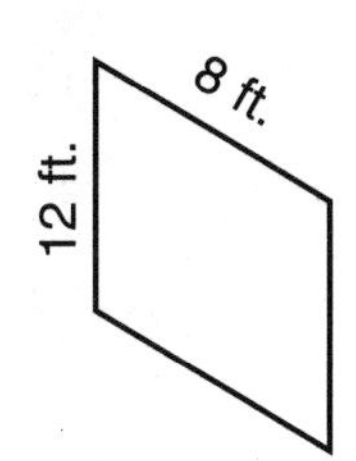

____

2.

15 cm
6 cm

____

3.

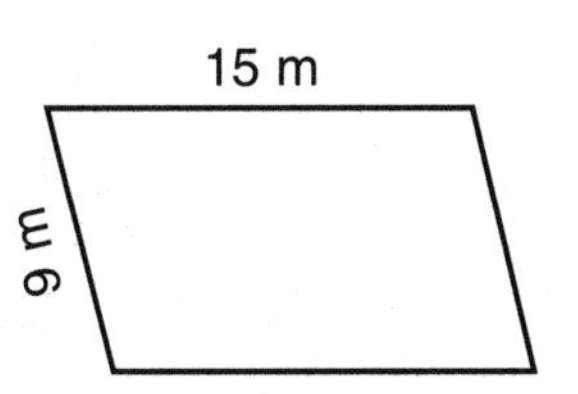

____

4.

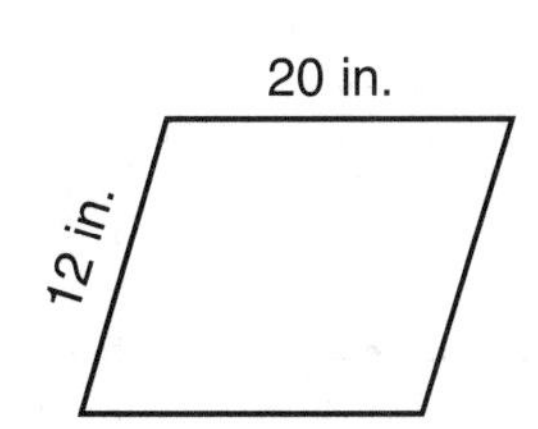

____

5.

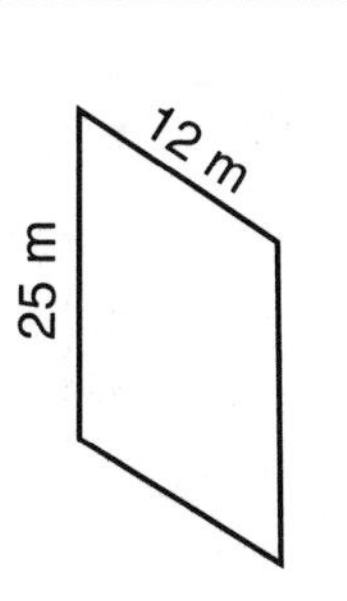

____

6.

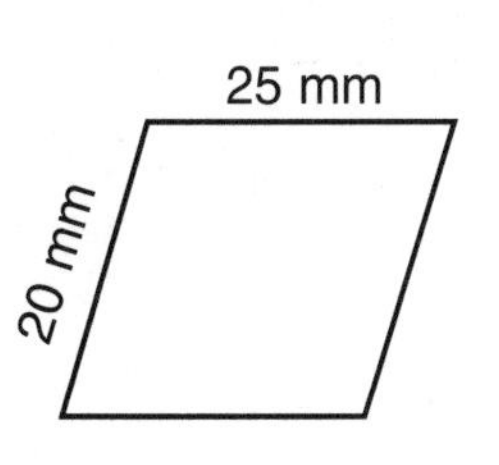

____

7.

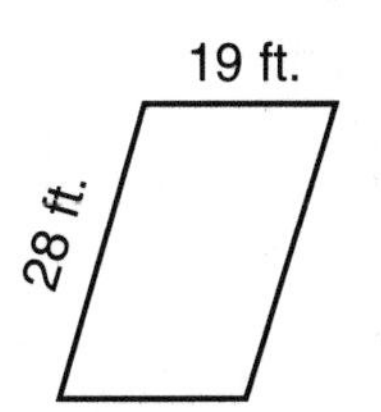

____

8.

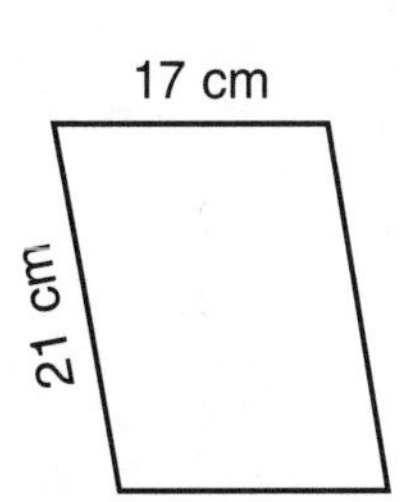

____

Practice 13

Reminder

The perimeter of figures with four unequal sides is computed by adding the lengths of each side.

Directions: Compute the perimeter of these parallelograms.

1.

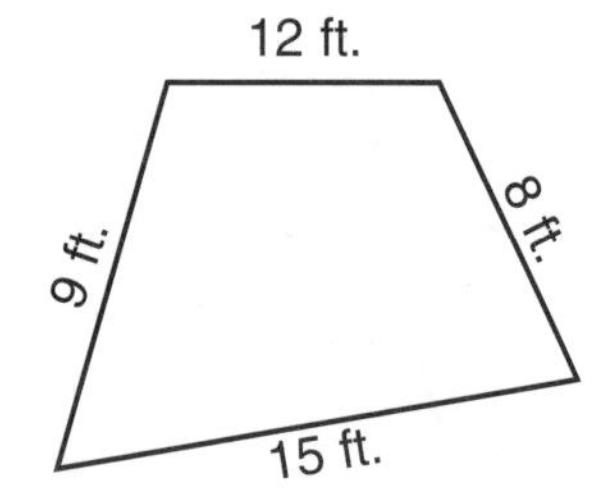

____

2.

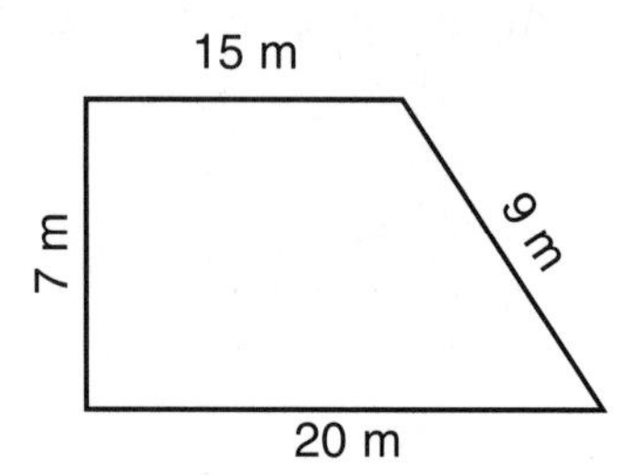

____

3

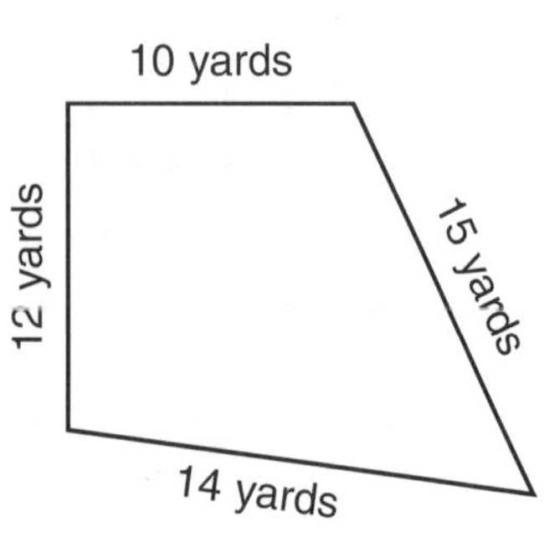

____

4.

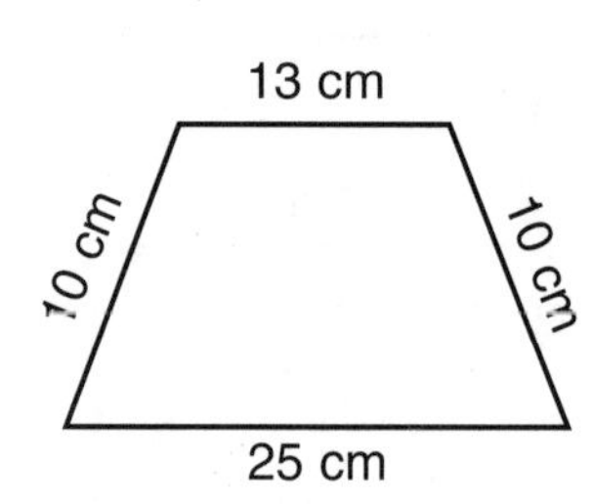

____

5.

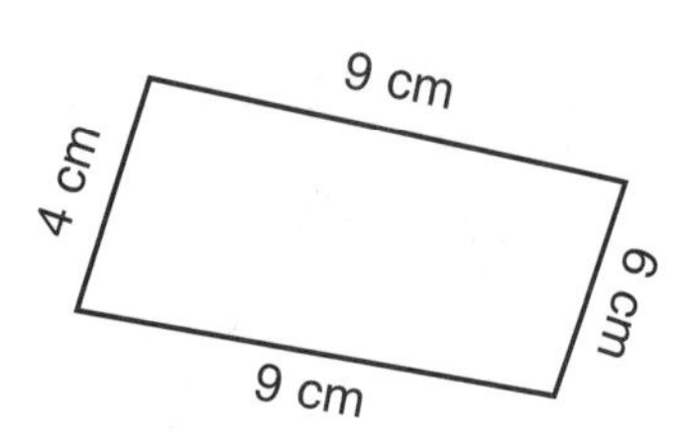

____

6.

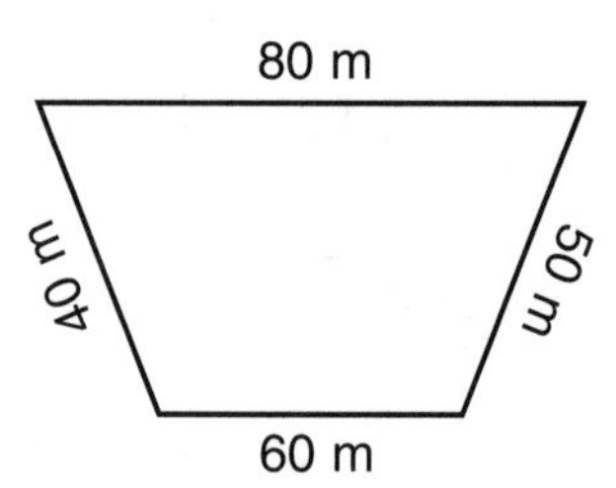

____

7.

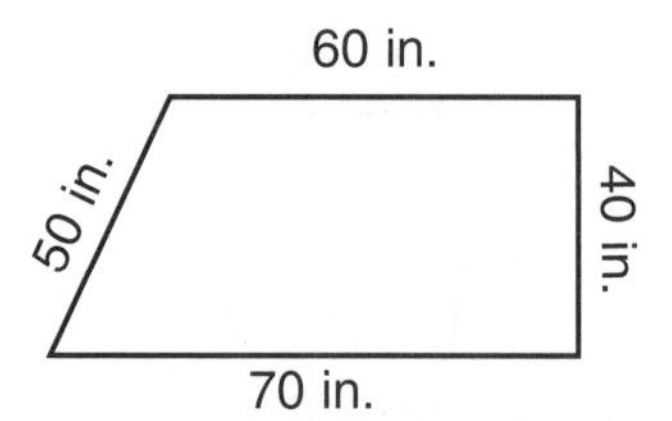

____

8.

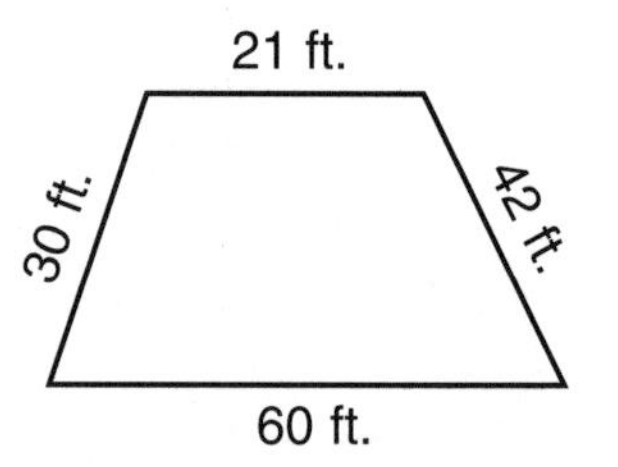

____

Practice 14

Reminders

- A regular polygon is a polygon which has equal sides and equal angles.
- The perimeter of a regular polygon is computed by multiplying the length of one side by the number of sides.

Directions: Compute the perimeter of each regular polygon.

1.

____

2.
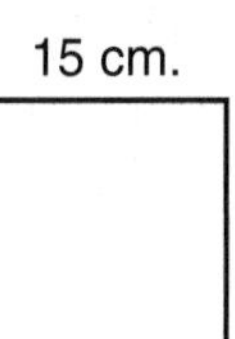

____

3.
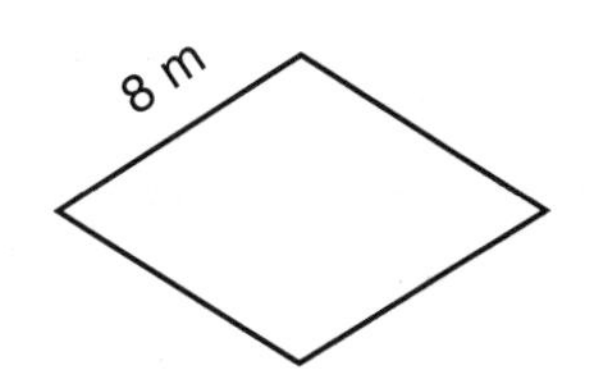

____

4.
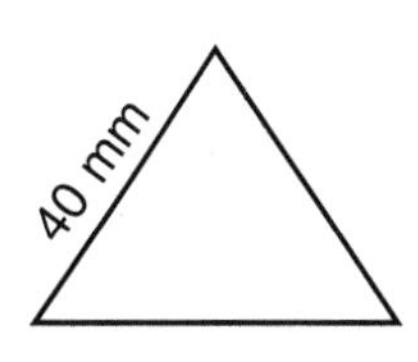

____

5.
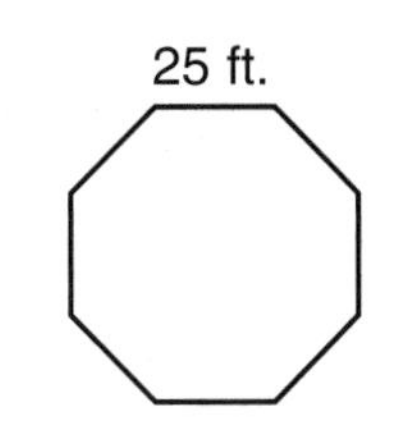

____

6.
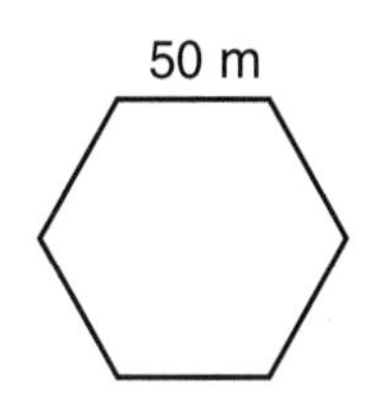

____

7.
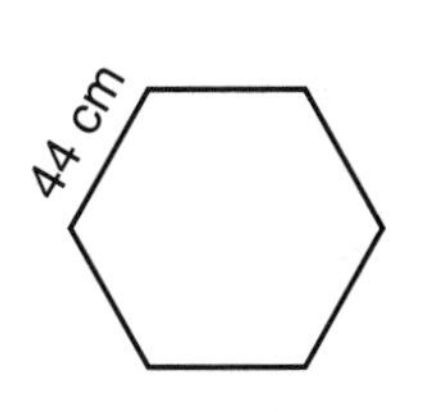

____

8.
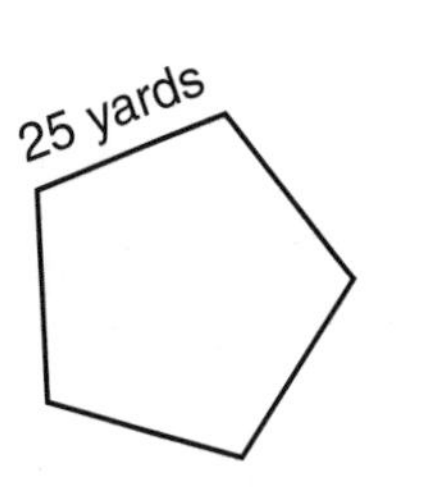

____

9.
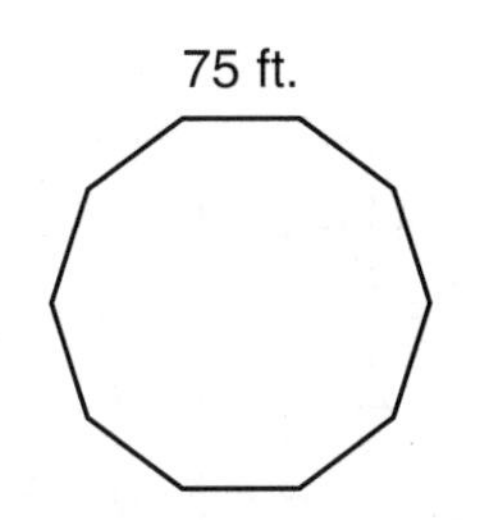

____

10.

____

Practice 15

Reminders

- The area of a flat surface is a measure of how much space is covered by that surface.
- Area is measured in square units.
- The area of a rectangle equals length times width.

A = l x w

Directions: Compute the number of square centimeters in these figures.

1. ___cm²

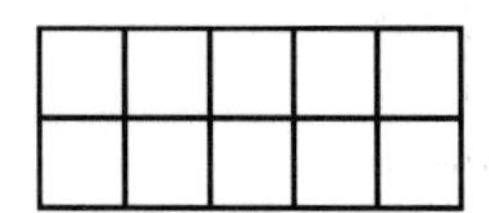

2. ______ cm²

3. ______ cm²

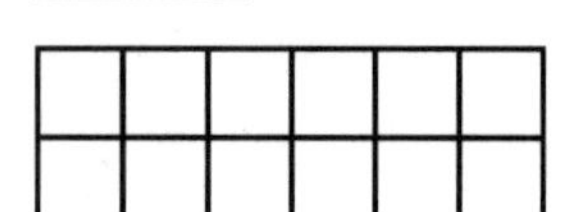

4. ______ cm²

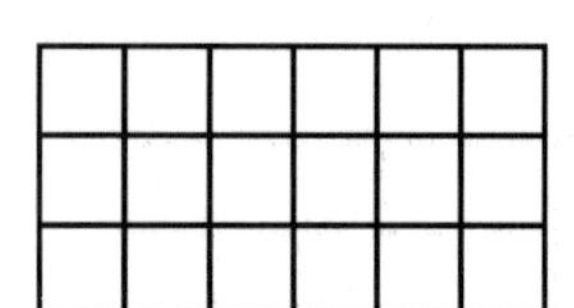

5. ______ cm²

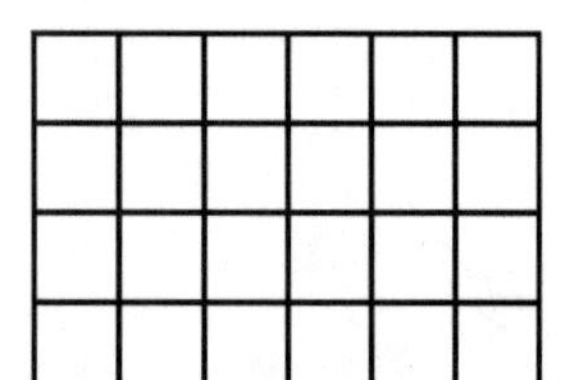

6. ______ cm²

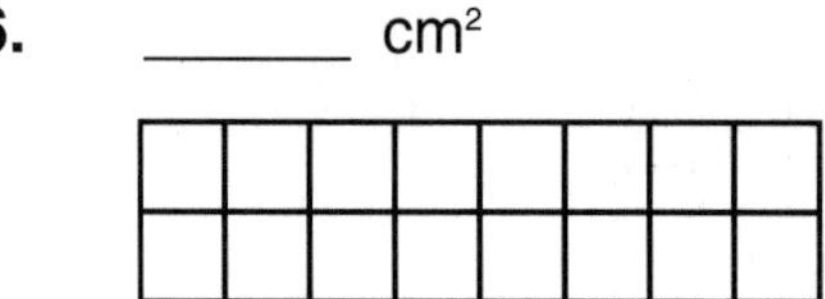

7. ______ cm²

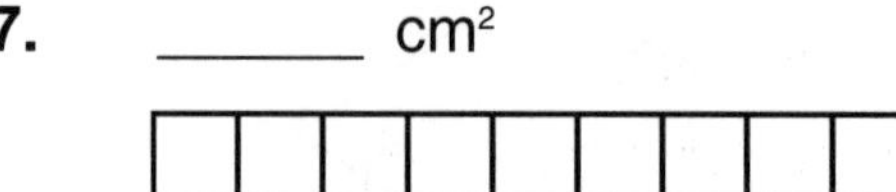

8. ______ cm²

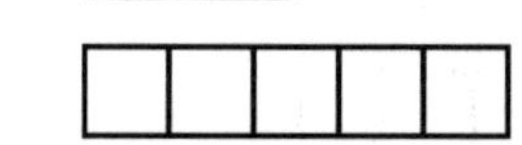

9. ______ cm²

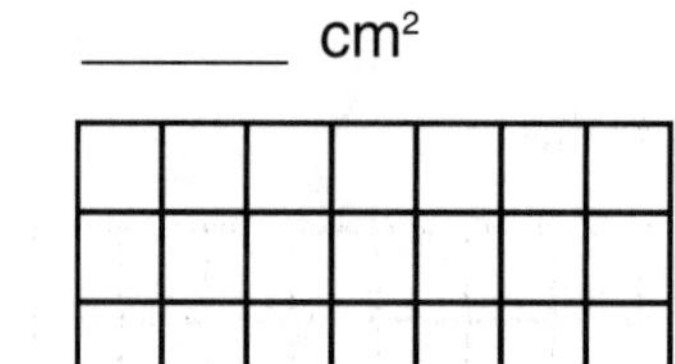

10. ______ cm²

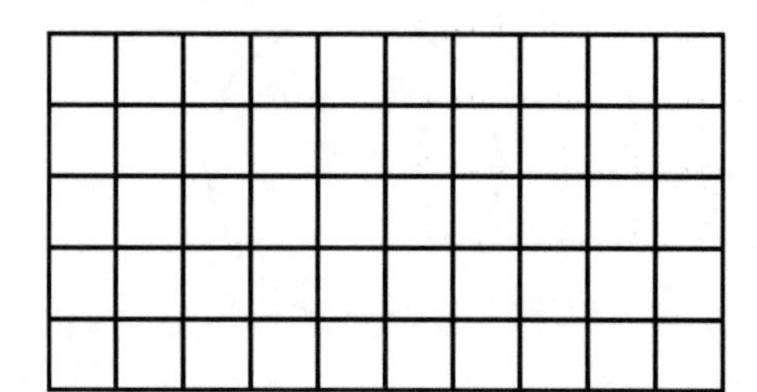

Practice 16

Reminder

The area of a rectangle is computed by multiplying the length times the width.

A = l x w

Directions: Count the number of squares along the width of the rectangle. Count the number of squares along the length of the rectangle. Multiply the length times the width to compute the area of each rectangle.

1. ______ sq. units

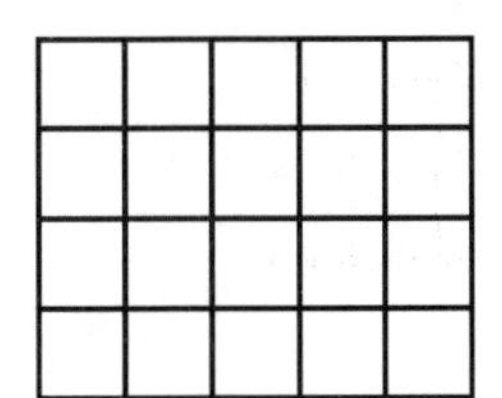

2. ______ sq. units

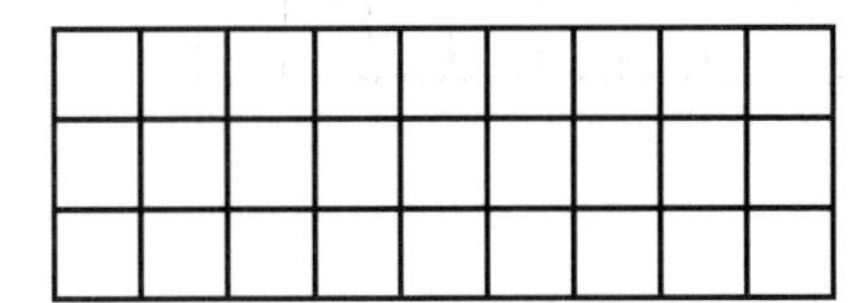

3. ______ sq. units

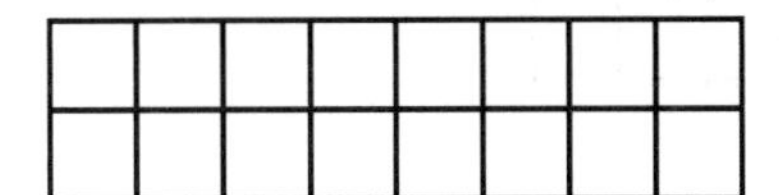

4. ______ sq. units

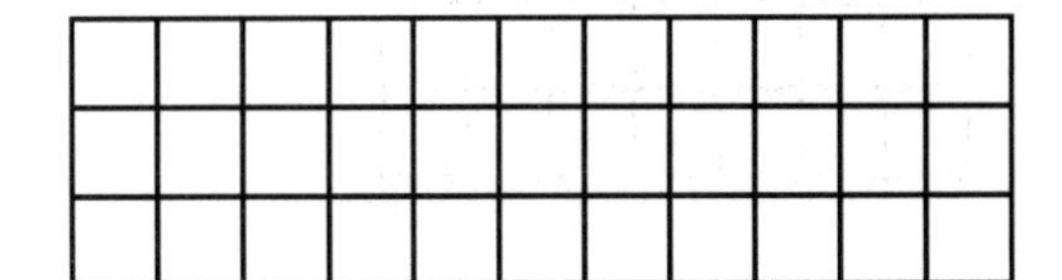

5. ______ sq. units

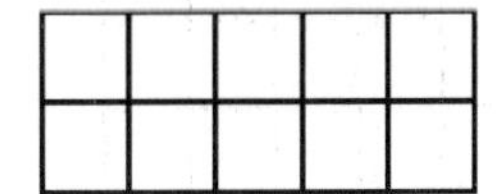

6. ______ sq. units

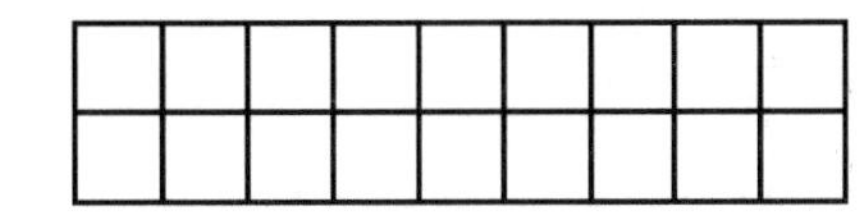

7. ______ sq. units

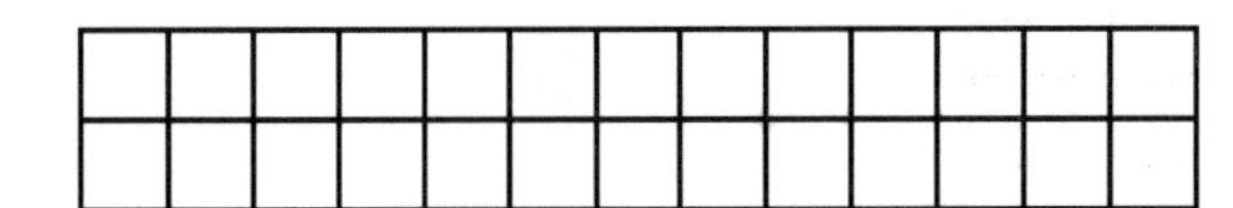

8. ______ sq. units

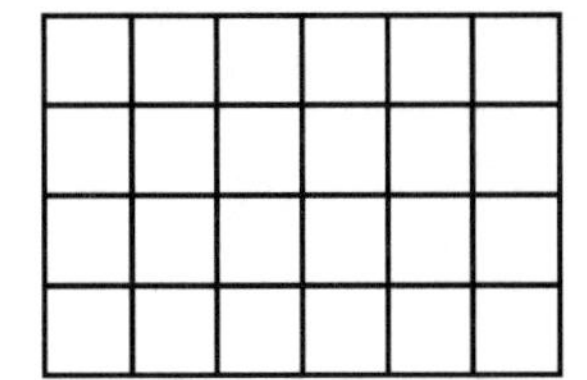

9. ______ sq. units

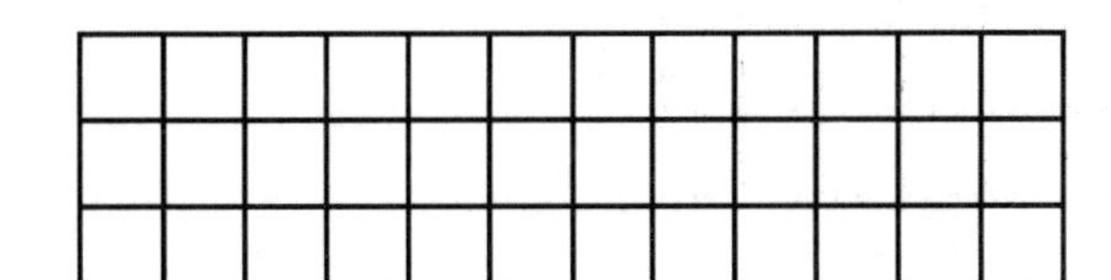

10. ______ sq. units

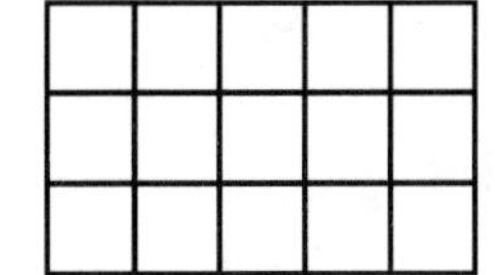

Practice 17

Reminder

The area of a square is computed by multiplying the length of one side times itself.

$A = s \times s$ or $A = s^2$ (Area = side squared)

Directions: Compute the area of each square.

1. 8 ft

____

2. 20 ft.

____

3. 15 m

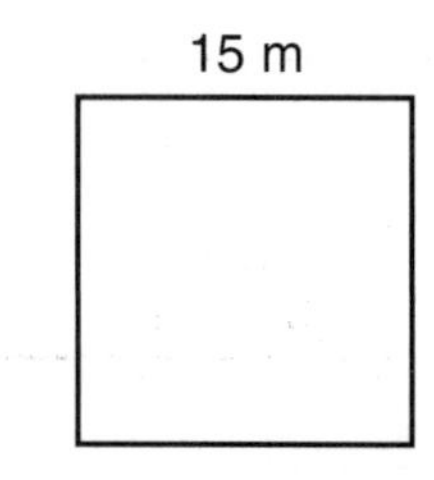

____

4. 25 cm

____

5. 18 yards

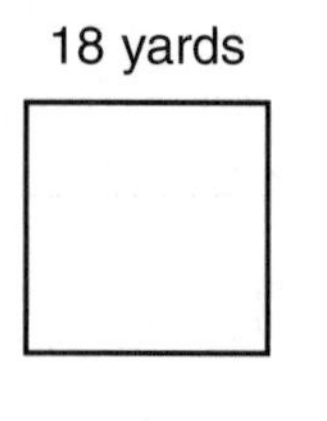

____

6. 11 cm

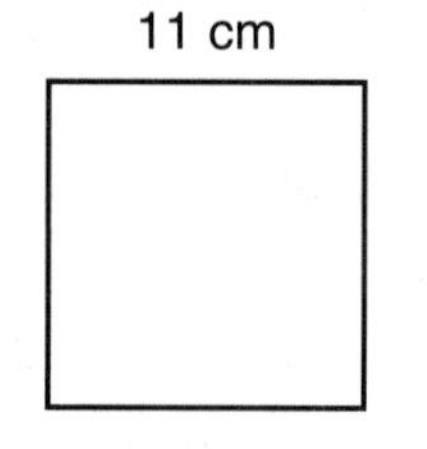

____

7. 62 ft.

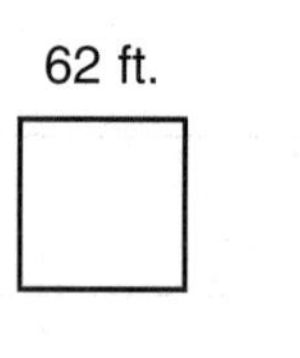

____

8. 11 miles

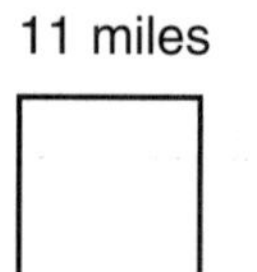

____

9. 2.2 m

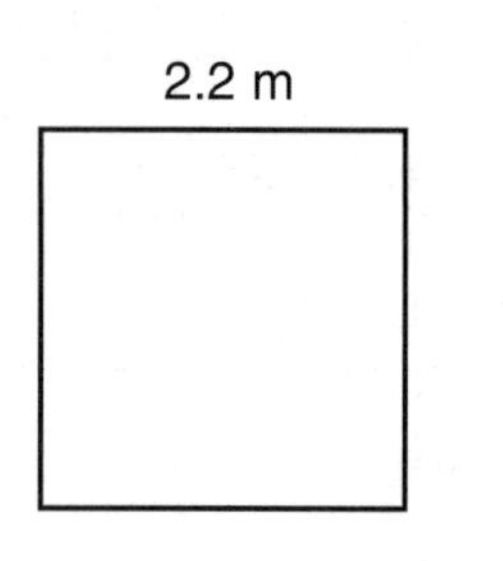

____

10. 1/2 ft.

____

Practice 18

Reminder

The area of a rectangle is computed by multiplying the width of one side times the length of the adjoining side.

$$A = l \times w$$

Directions: Compute the area of each rectangle.

1.

10 in.
7 in.

____

2.

12 cm
6 cm

____

3.

15 m
12 m

____

4.

40 cm
25 cm

____

5.

30 yards
11 yards

____

6.

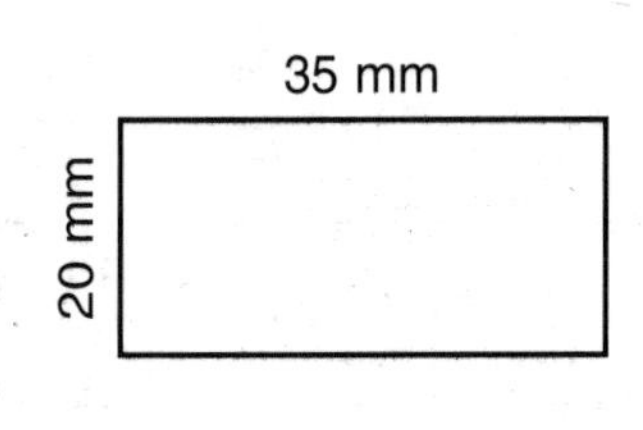

____

7.

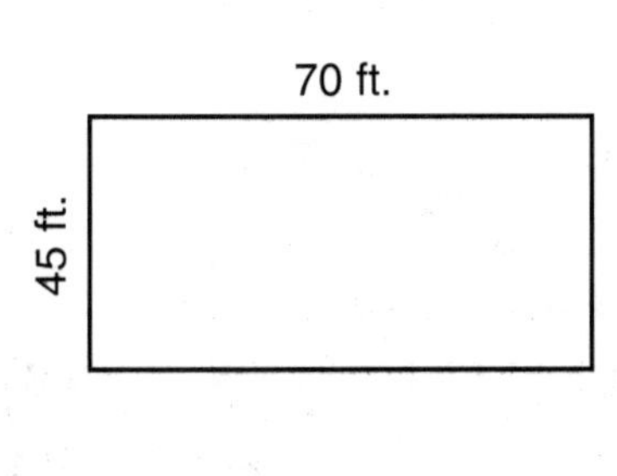

____

8.

200 mm
25 mm

____

Practice 19

Reminder

The area of a rectangle is computed by multiplying the width of one side times the length of the adjoining side.

A = l x w

Directions: Compute the area of each rectangle.

1.

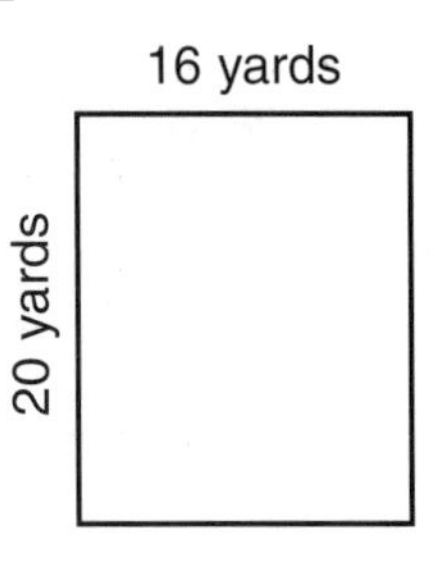

____

2.

23 mm

40 mm

____

3.

57 m

40 m

____

4.

93 ft.

60 ft.

____

5.

69 in.

13 in.

____

6.

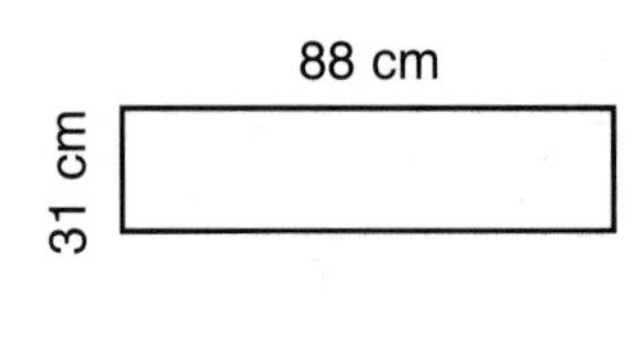

____

7.

59 m

17 m

____

8.

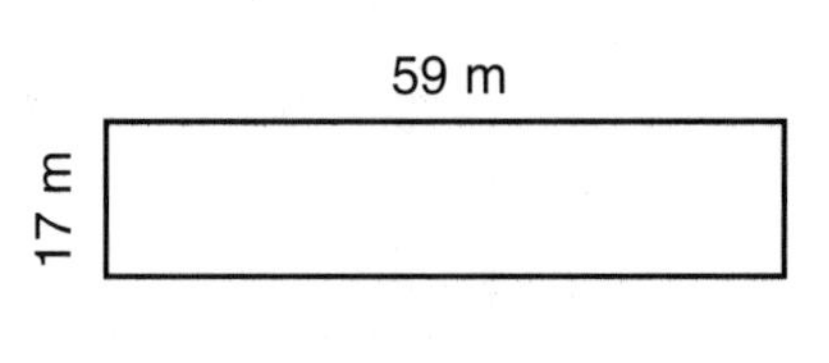

____

Practice 20

Reminder

The area of a parallelogram is computed by multiplying the base times the height: **A = b x h**

Directions: Compute the area of each parallelogram.

1.

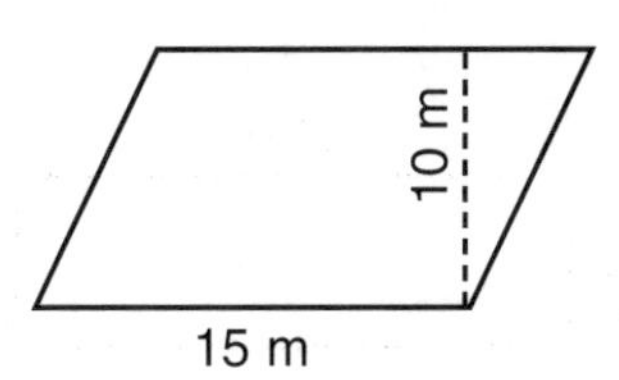

____

2.

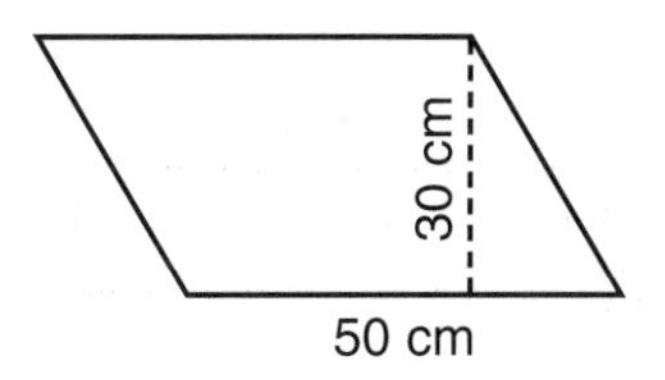

____

3.

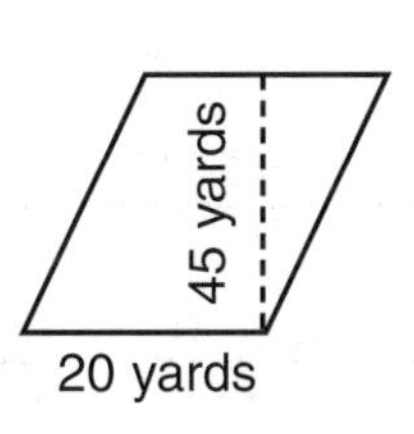

____

4.

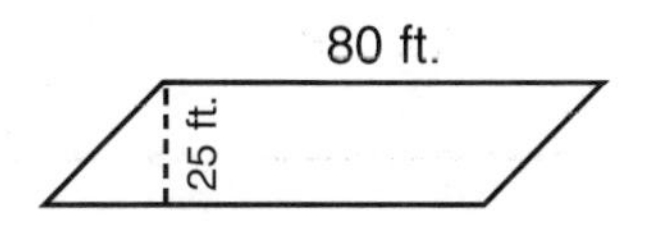

____

5.

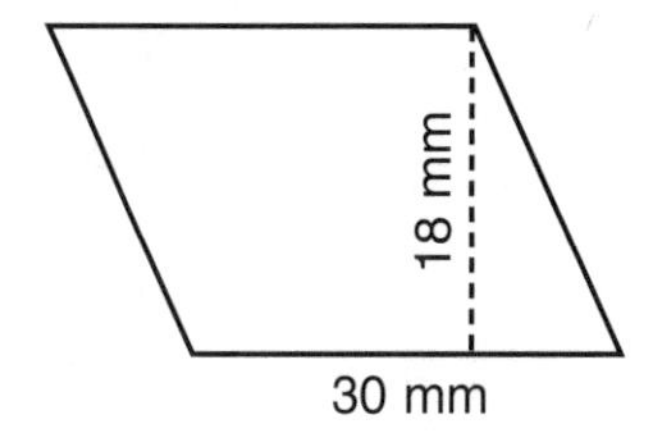

____

6.

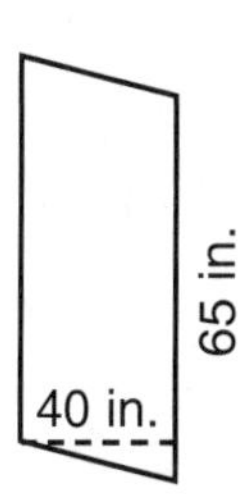

____

7.

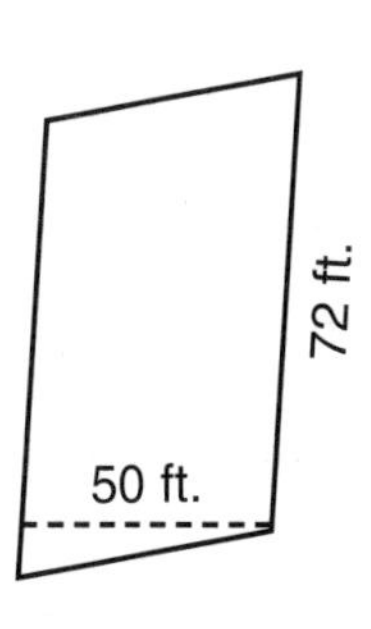

____

8.

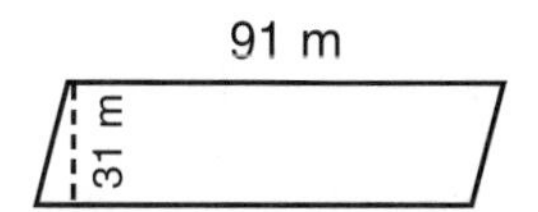

____

Practice 21            

Reminder

The area of a parallelogram is computed by multiplying the base times the height: **A = b x h**

Directions: Compute the area of each parallelogram.

1.

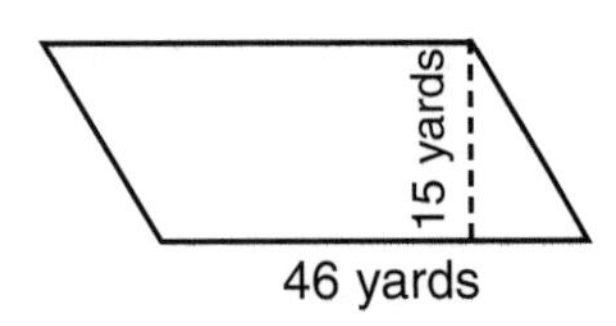

____

2.

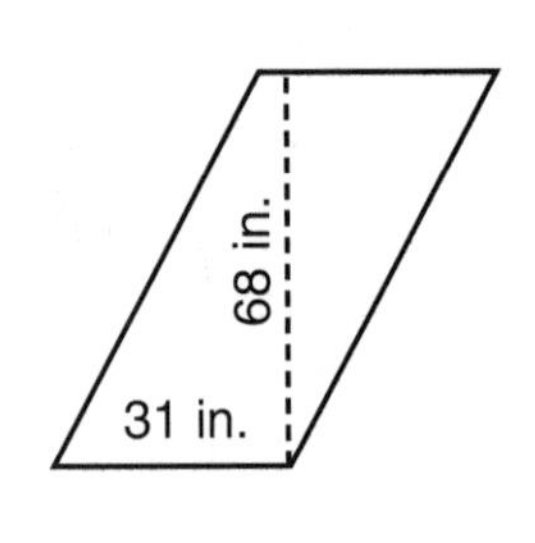

____

3.

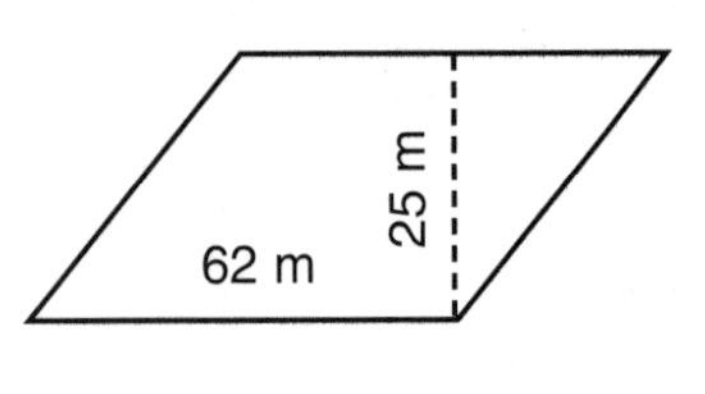

____

4.

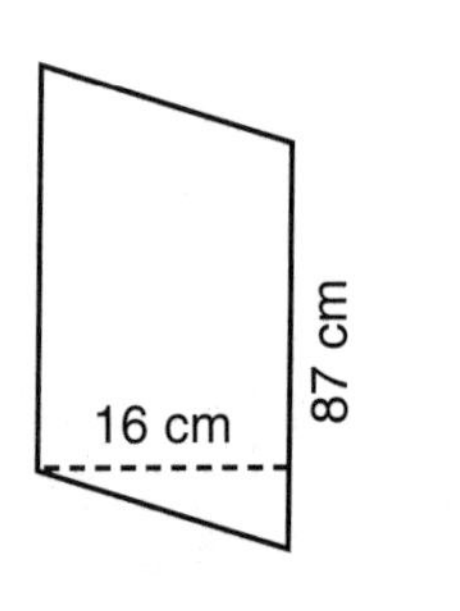

____

5.

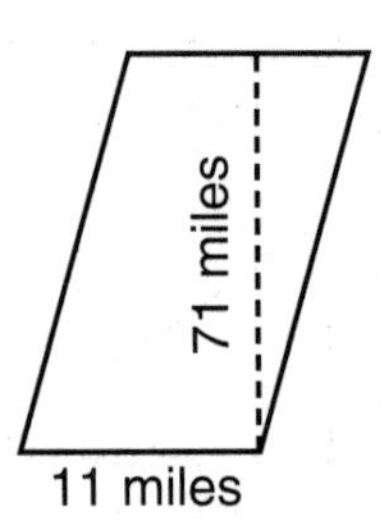

____

6.

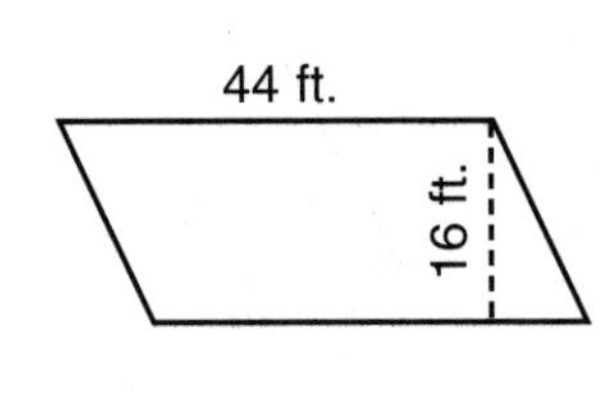

____

7.

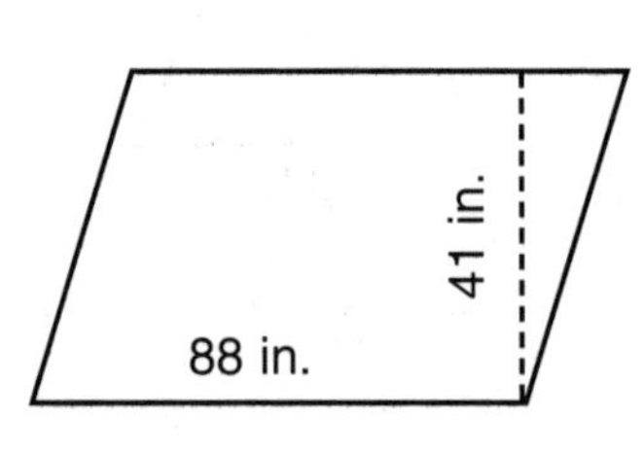

____

8.

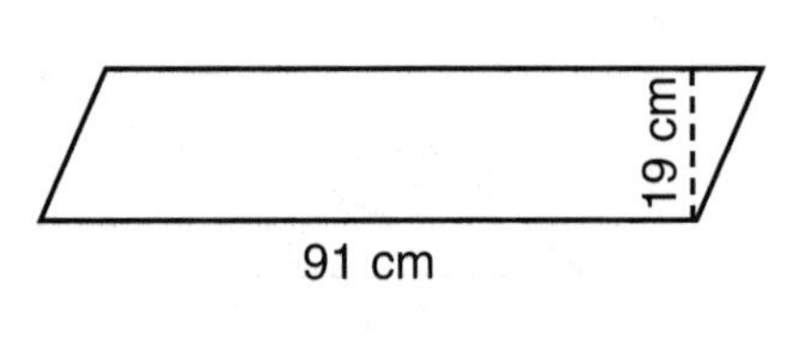

____

Practice 22

Reminders

- The area of a triangle is one half the area of a parallelogram.
- To compute the area of a triangle, multiply the base times the height and divide by 2 or multiply 1/2 by the base times the height.

A = 1/2 (b x h)

Directions: Compute the area of each triangle.

1.

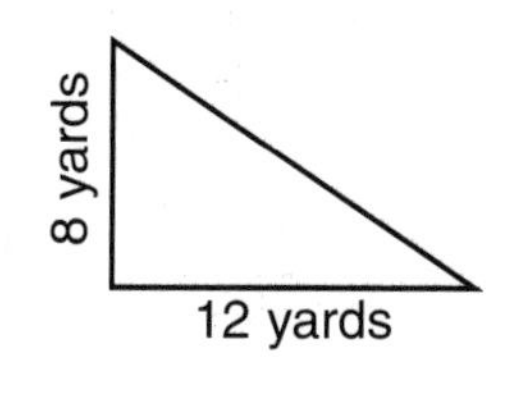

____

2.

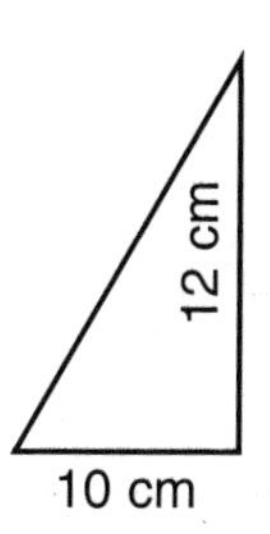

____

3.

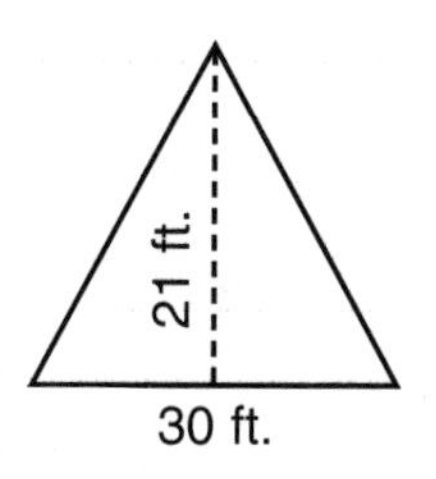

____

4.

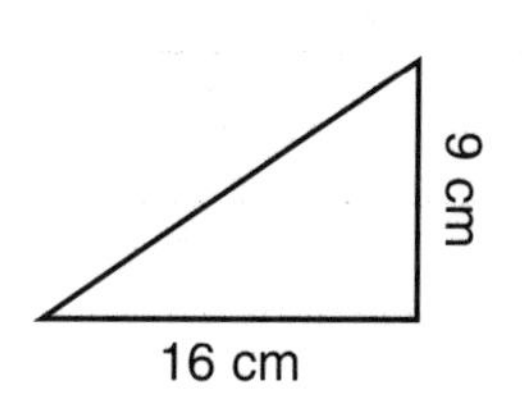

____

5.

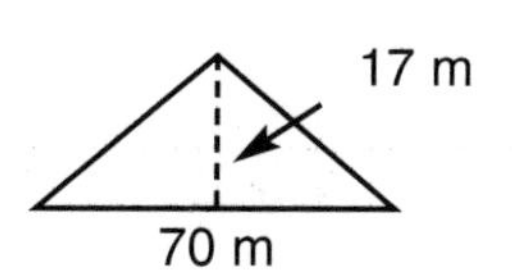

____

6.

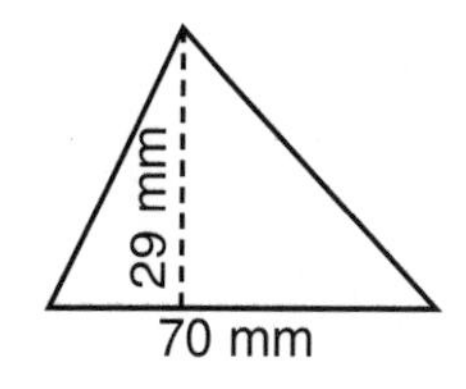

____

7.

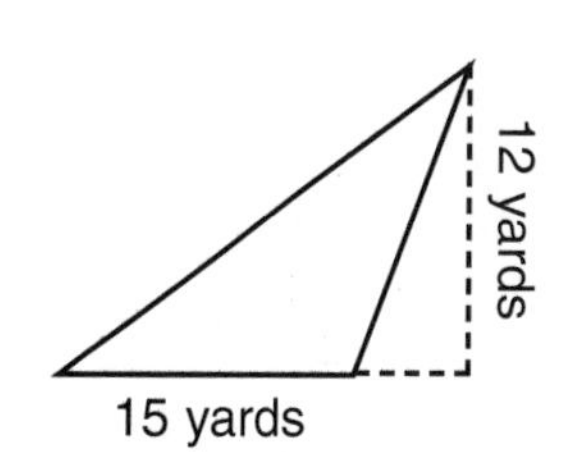

____

8.

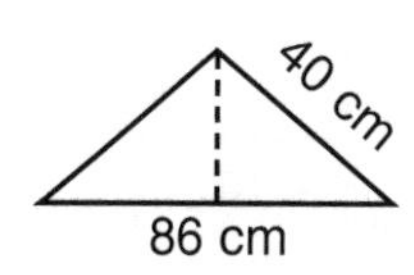

____

9.

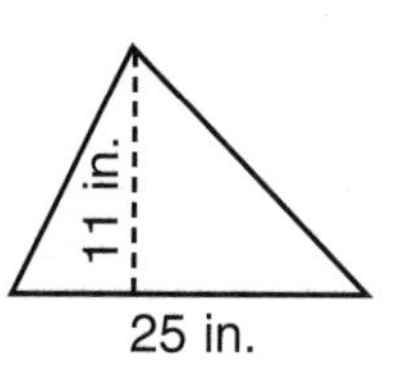

____

10.

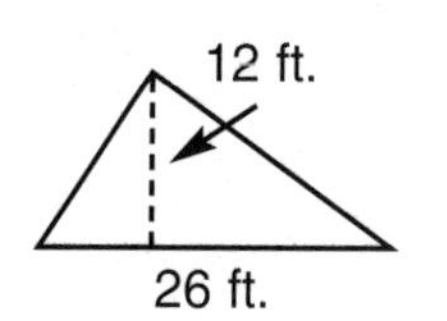

____

Practice 23

Reminders

- The area of a triangle is one half the area of a parallelogram.
- To compute the area of a triangle, multiply the base times the height and divide by 2 or multiply 1/2 by the base times the height.

A = 1/2 (b x h)

Directions: Compute the area of each triangle.

1.

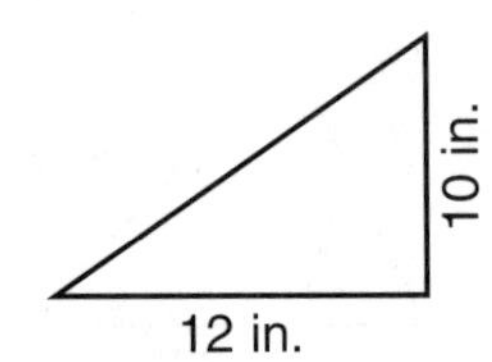

____

2.

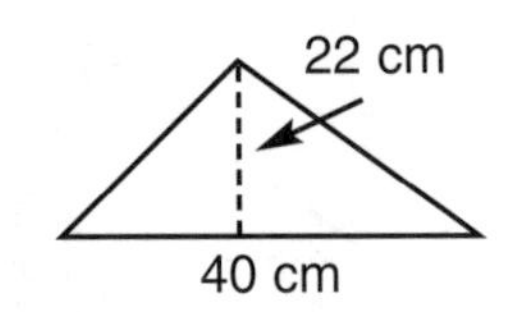

____

3.

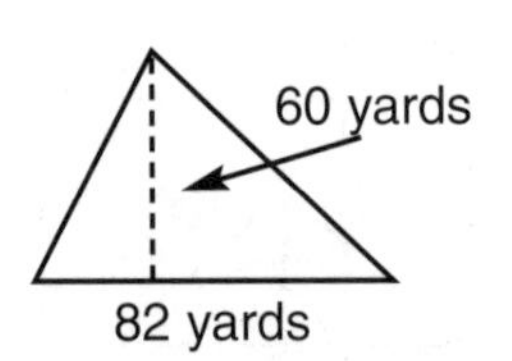

____

4.

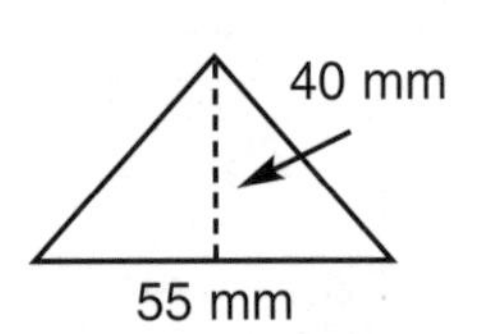

____

5.

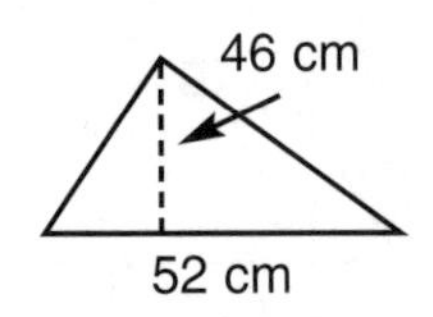

____

6.

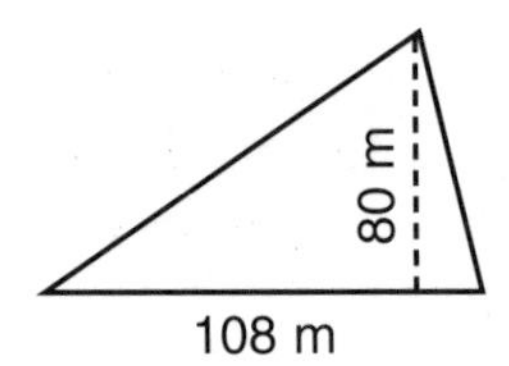

____

7.

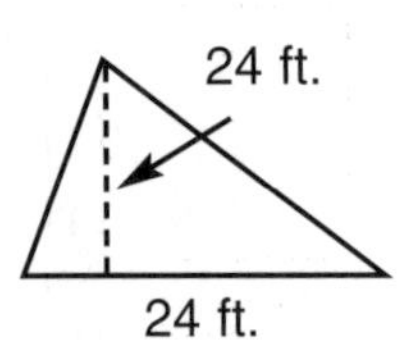

____

8.

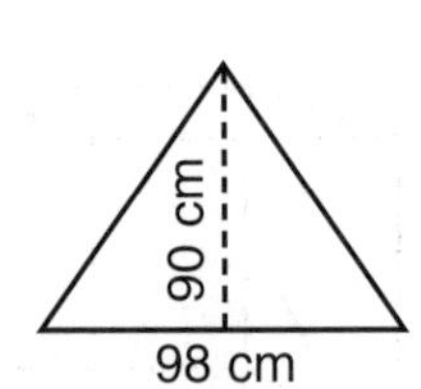

____

9.

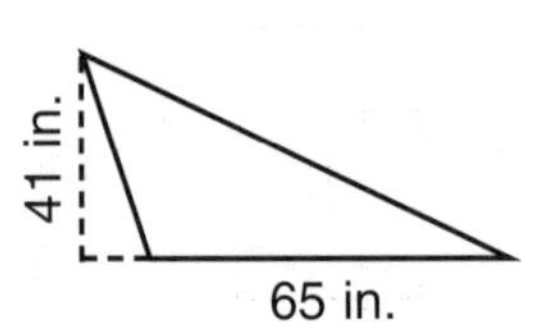

____

10.

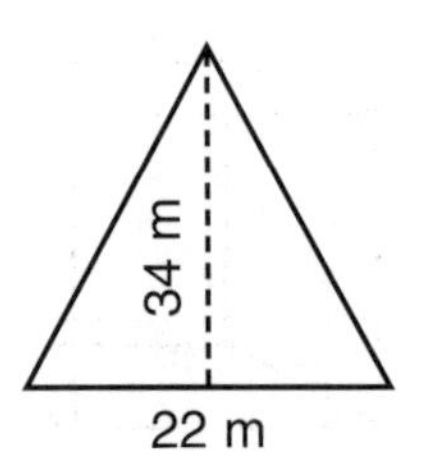

____

Practice 24

Reminder

The volume of a cube is computed by multiplying the length of one side times itself times itself again.

V = s x s x s or V = s³ or Volume = side cubed

Directions: Compute the volume of each cube.

1.

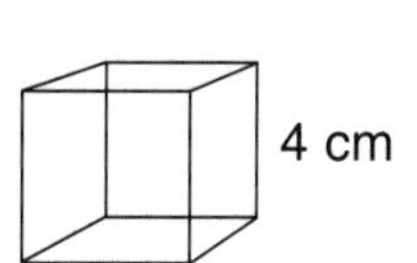

4 cm

____

6.

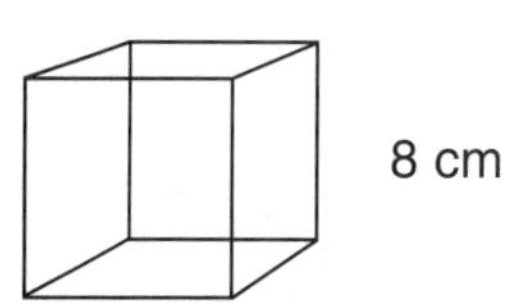

8 cm

____

2.

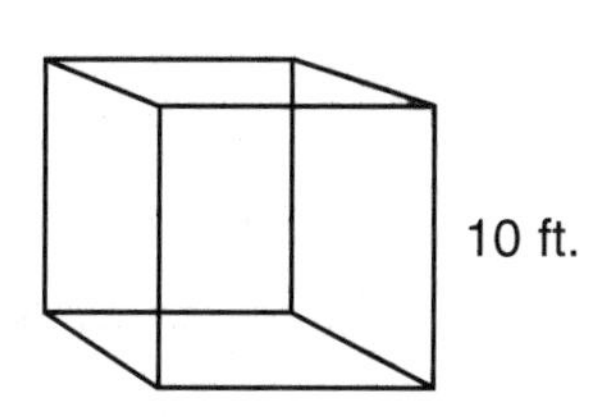

10 ft.

____

17.

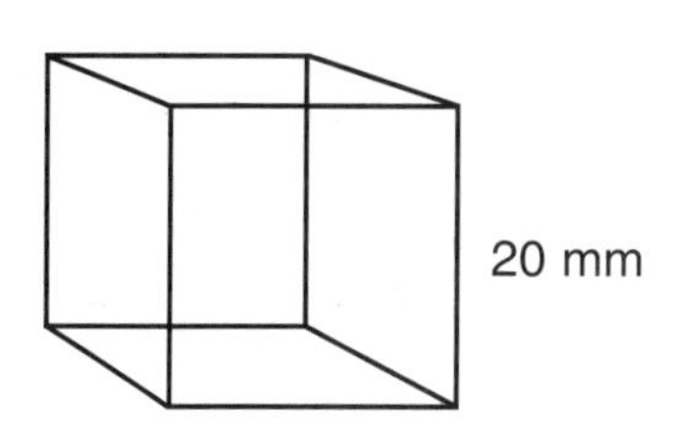

20 mm

____

3.

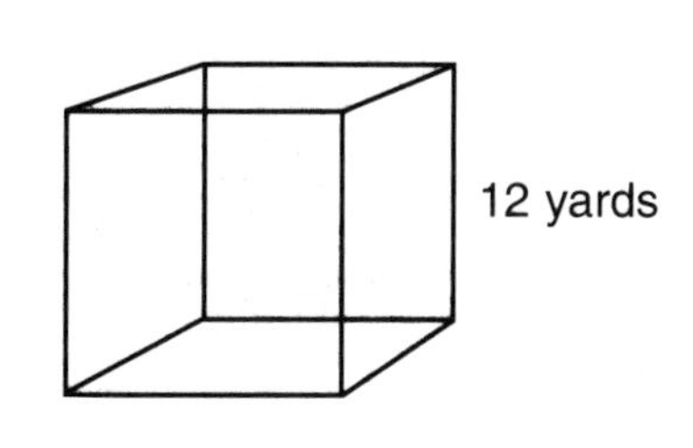

12 yards

____

18.

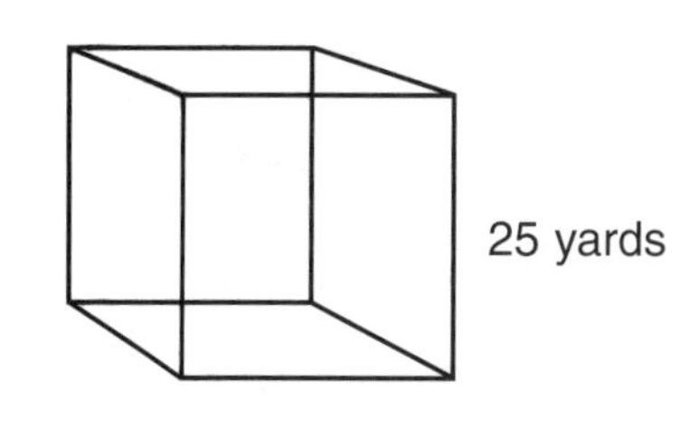

25 yards

____

4.

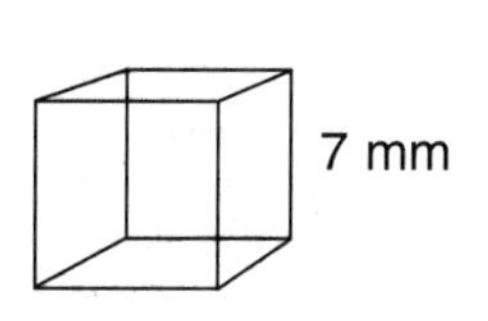

7 mm

____

9.

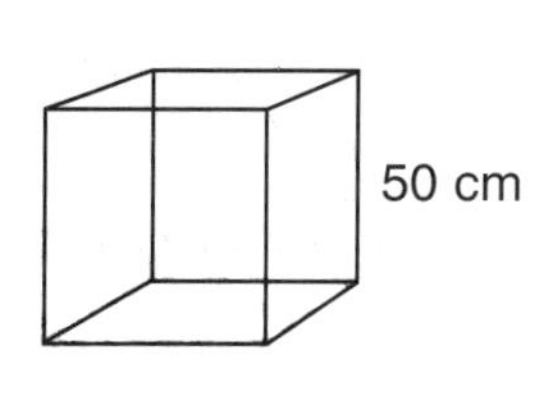

50 cm

____

5.

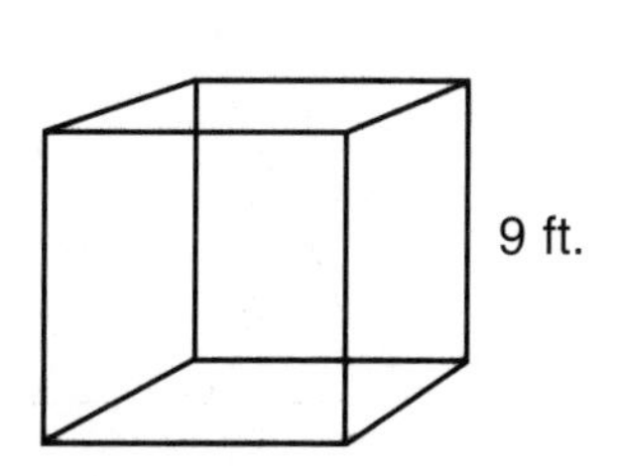

9 ft.

____

10.

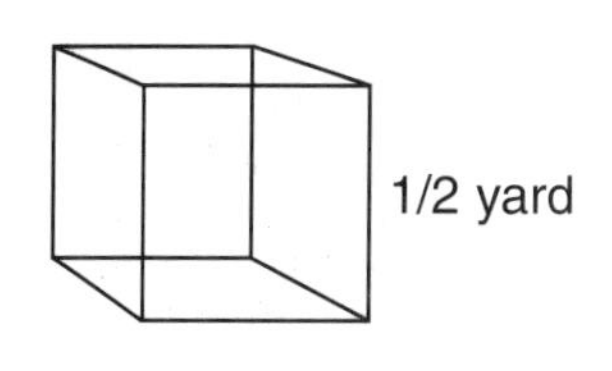

1/2 yard

____

Practice 25

Reminder

The volume of a rectangular prism is computed by multiplying the length times the width times the height of the prism.

$$V = l \times w \times h \text{ or } V = lwh$$

Directions: Compute the volume of each rectangular prism.

1.

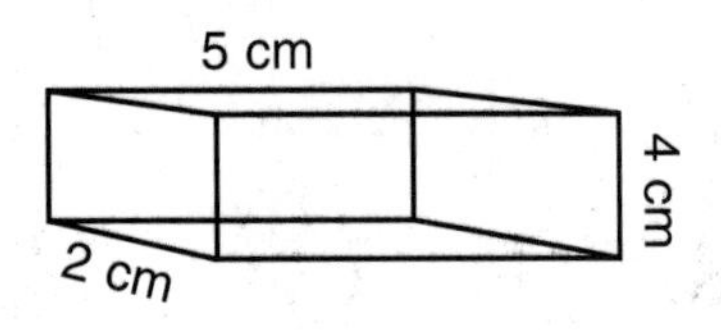

____

5.

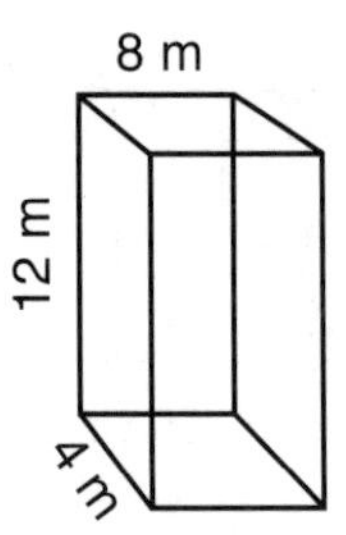

____

2.

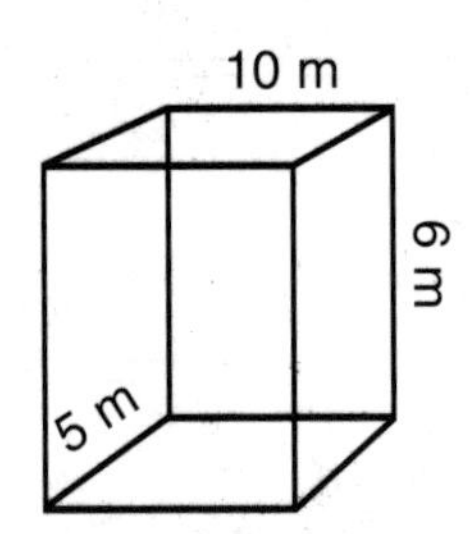

____

6.

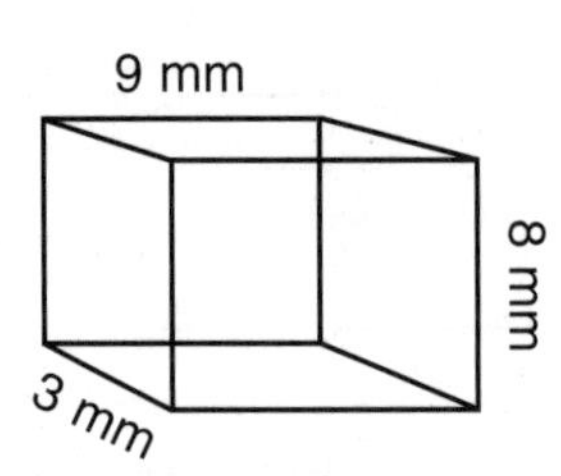

____

3.

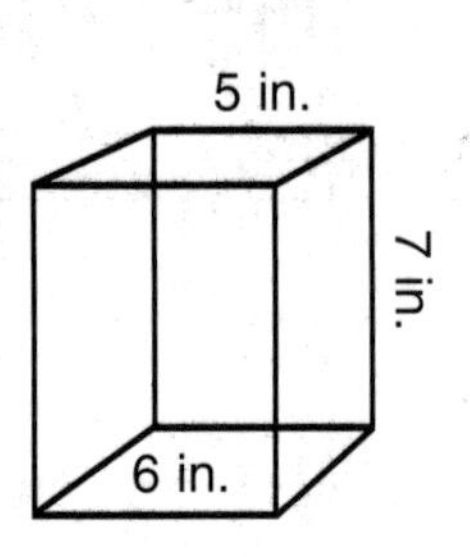

____

7.

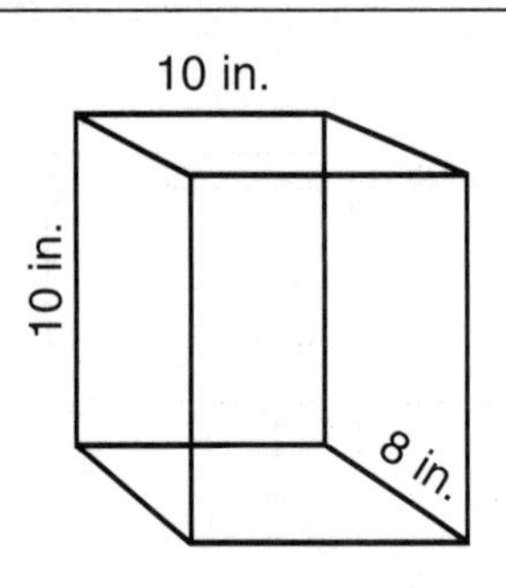

____

4.

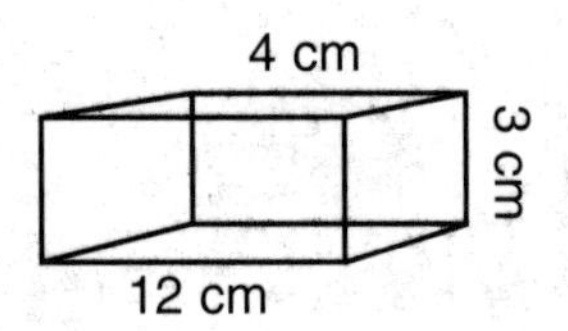

____

8.

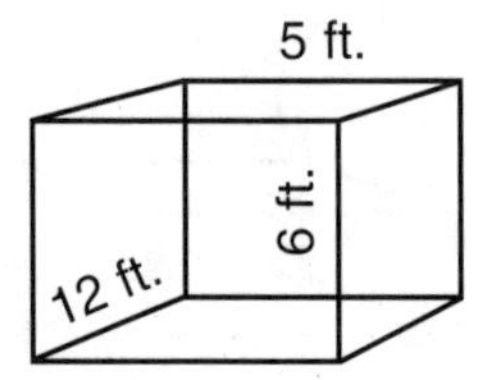

____

Practice 26

Reminder

The volume of a rectangular prism is computed by multiplying the length times the width times the height of the prism.

V = l x w x h or V = lwh

Directions: Compute the volume of each rectangular prism.

1.

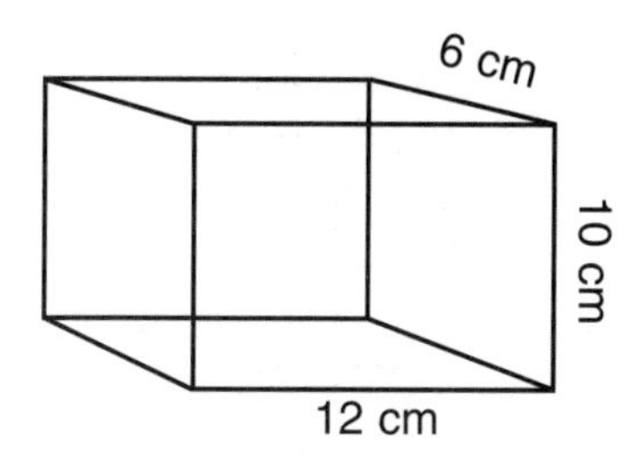

____

2.

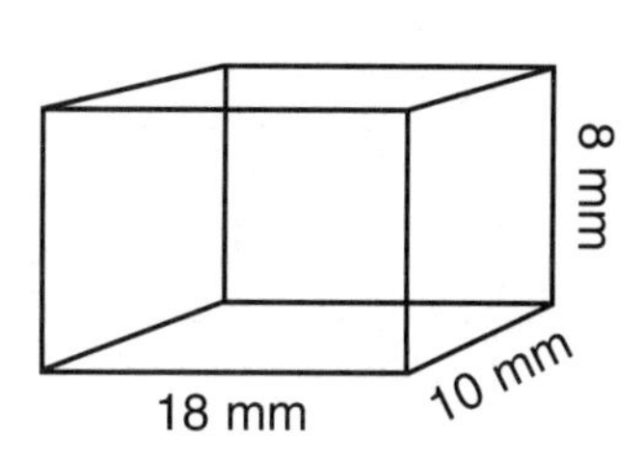

____

3.

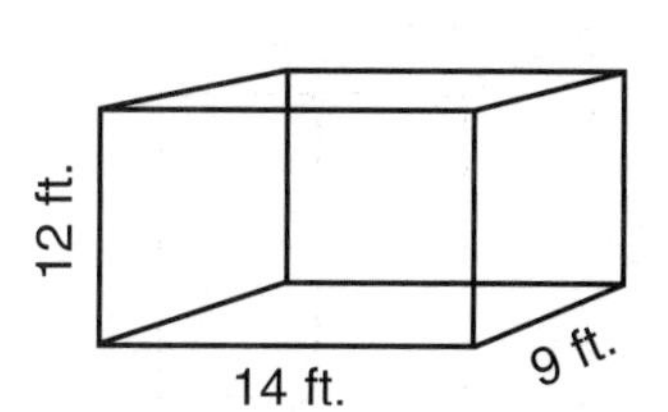

____

4.

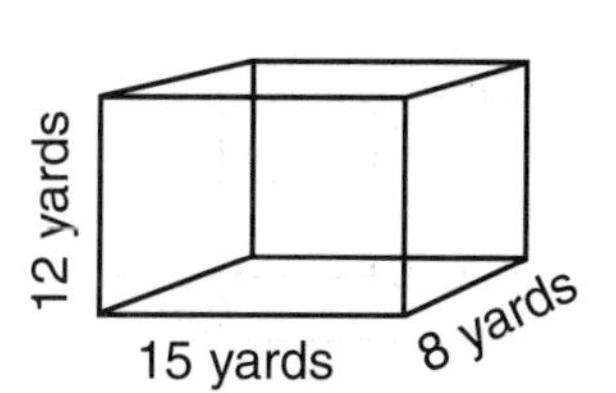

____

5.

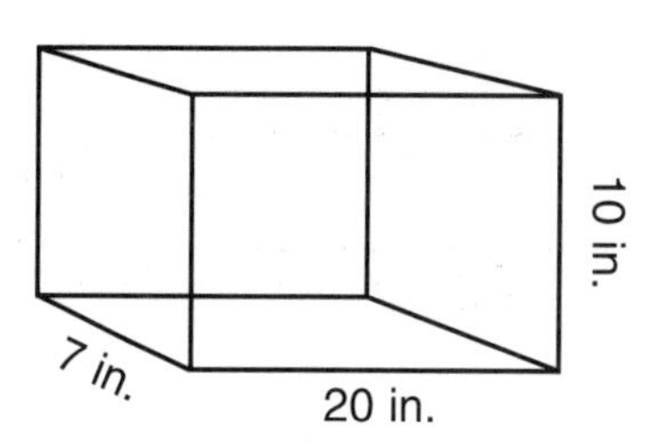

____

6.

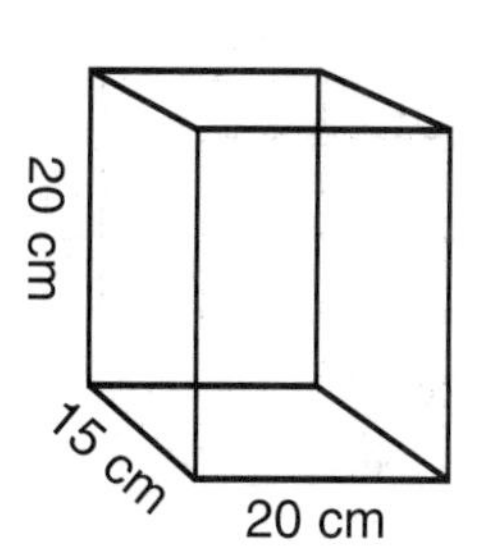

____

7.

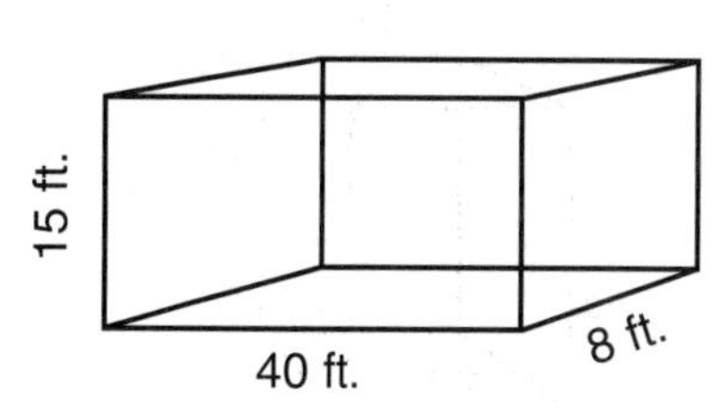

____

8.

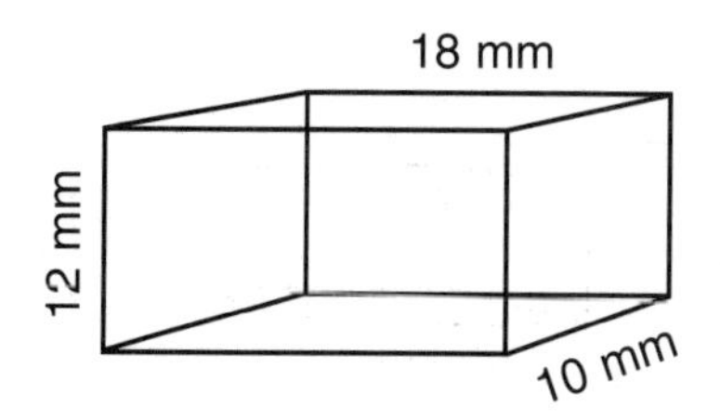

____

Practice 27

Reminders

- The circumference is the distance around a circle.
- The radius is the distance from the center of a circle to any point on the circle.
- The diameter is a line segment extending from one side of the circle to the other through the center of the circle.
- The diameter is twice the radius.

Directions: Label the circumference, the radius, and the diameter on these circles.

1.

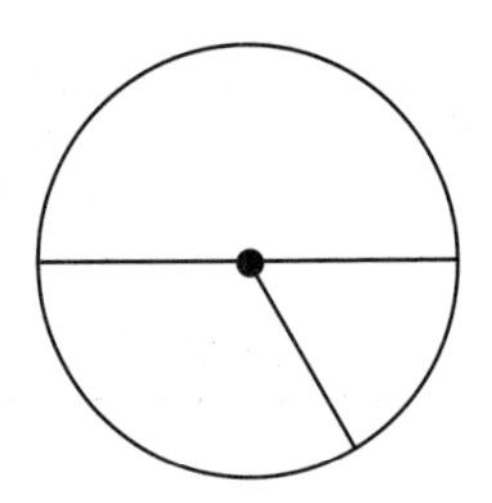

2.

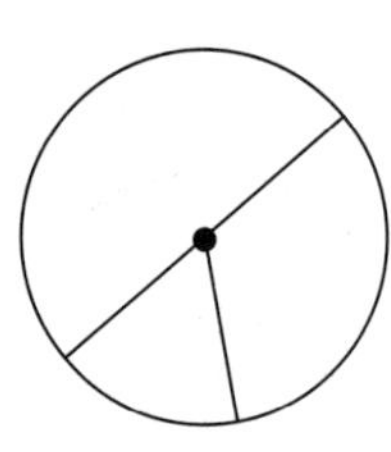

3.

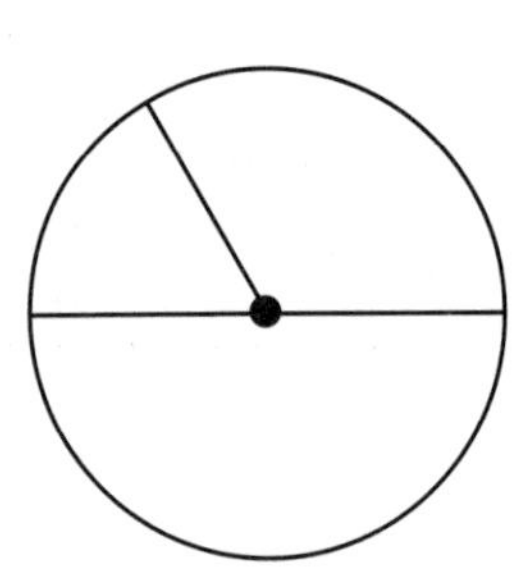

4.

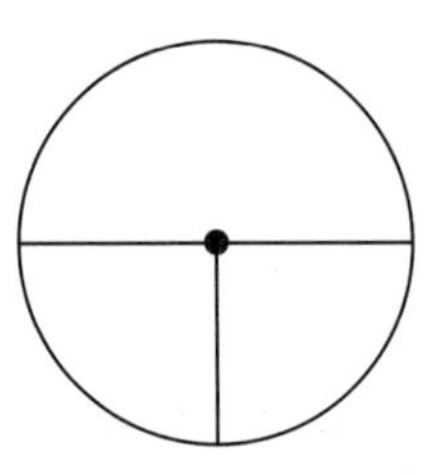

Directions: Use the information on the circles to find the values.

5.

radius = __________

diameter = __________

circumference = __________

14 ft.

6.

radius = __________

diameter = __________

circumference = __________

36 cm

7.

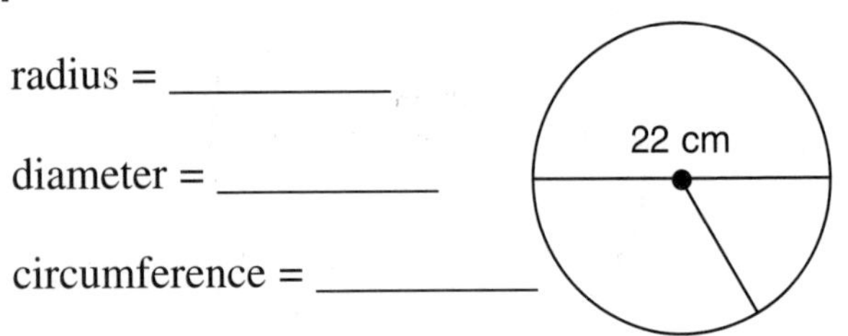

radius = __________

diameter = __________

circumference = __________

8.

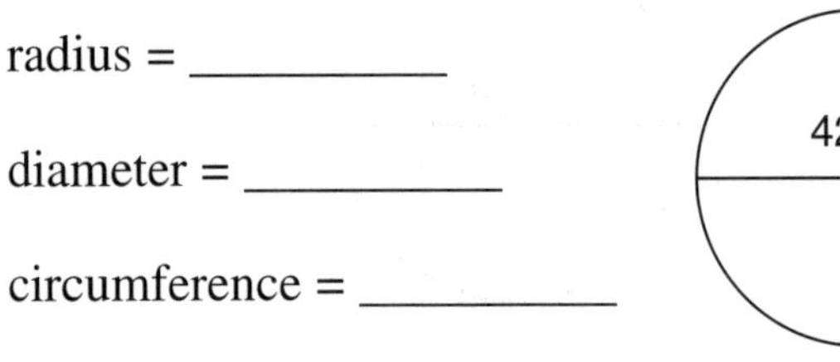

radius = __________

diameter = __________

circumference = __________

42 in.

Practice 28

  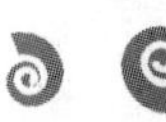        

Reminders

- The circumference is the distance around a circle.
- Pi = 3.14
- The circumference is computed by multiplying 3.14 times the diameter.

C = πd (Pi times the diameter)

Directions: Compute the circumference of each circle.

1.

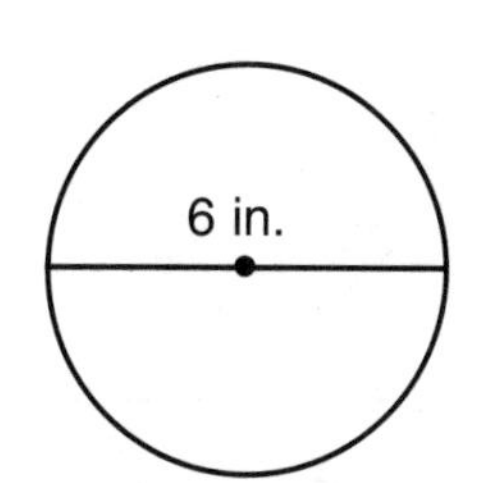

C = ____________________

2.

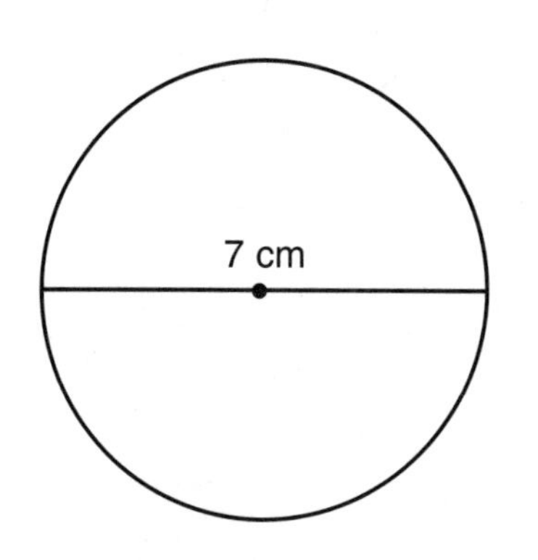

C = ____________________

3.

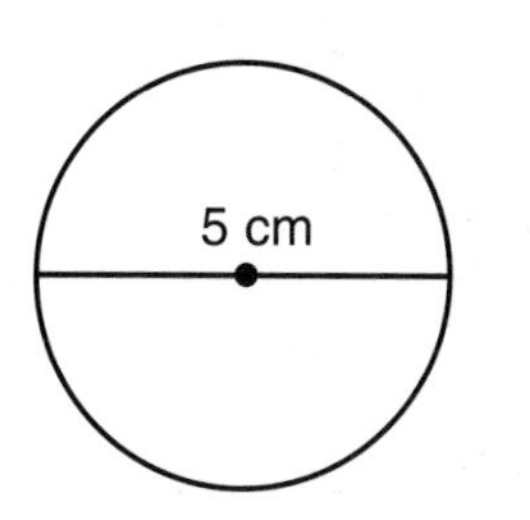

C = ____________________

4.

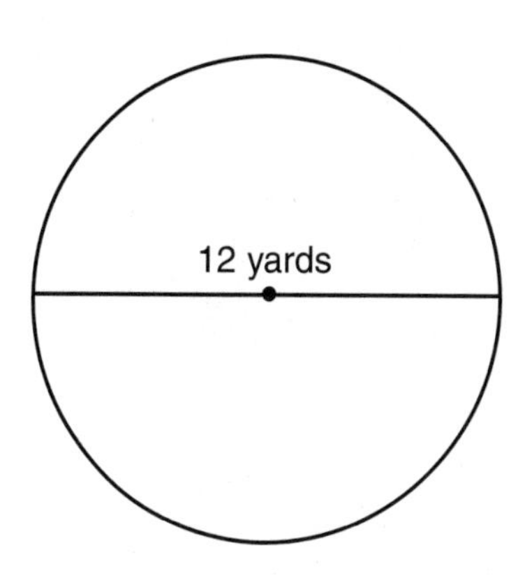

C = ____________________

5.

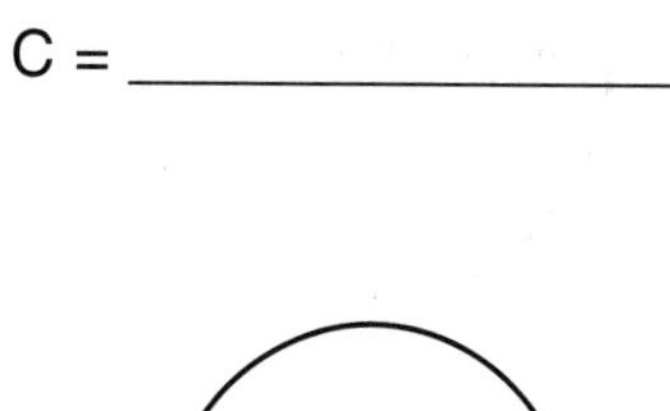
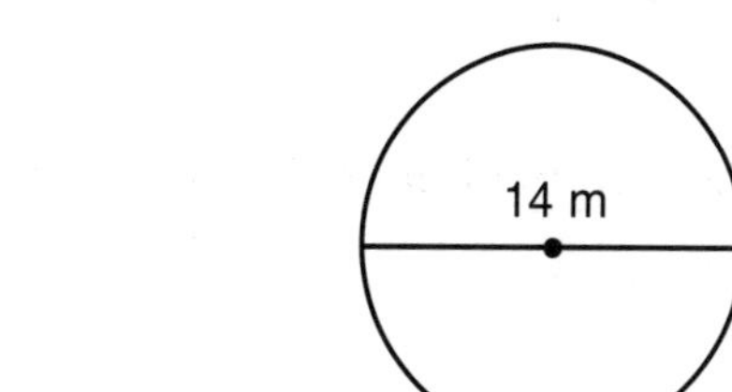

C = ____________________

6.

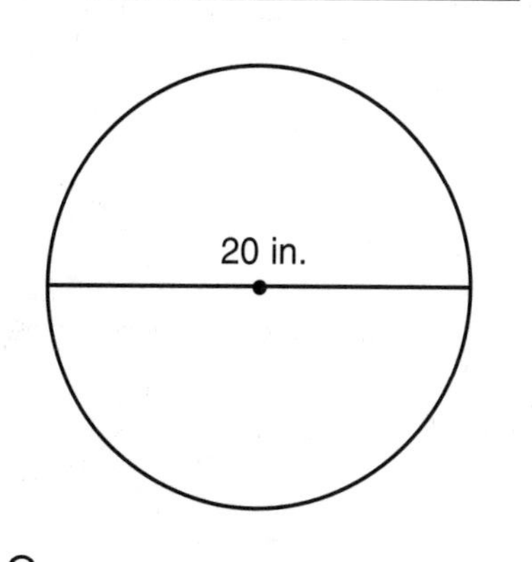

C = ____________________

Practice 29

Reminders

- The diameter is twice the radius.
- Pi = 3.14
- The area of a circle is computed by multiplying the radius times itself and by 3.14.

$A = r^2$ (Pi times the radius squared)

Directions: Compute the area of each circle.

1.

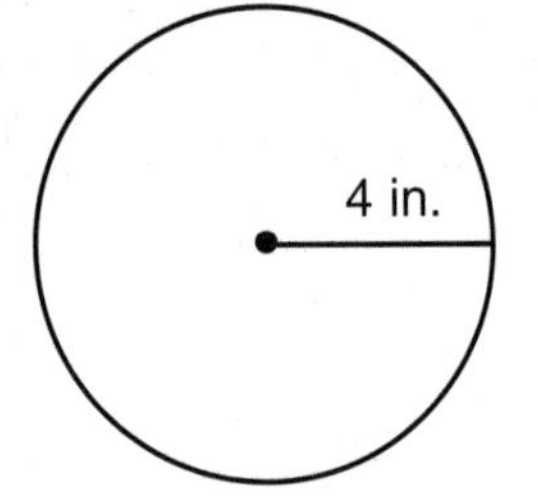

A = ____________________

2.

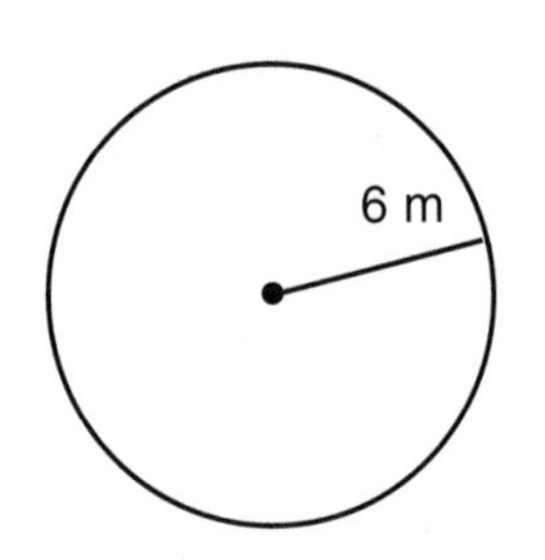

A = ____________________

3.

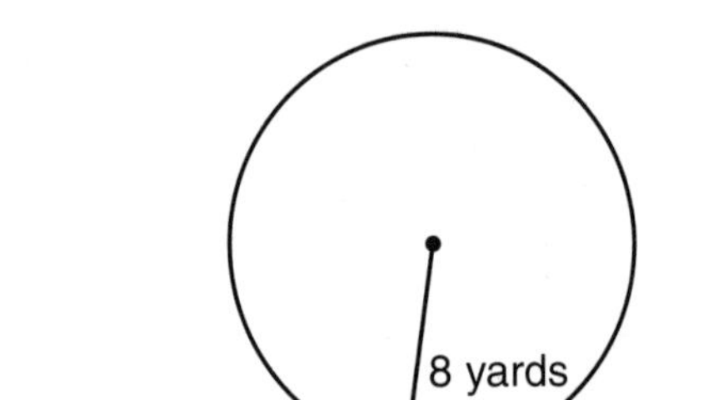

A = ____________________

4.

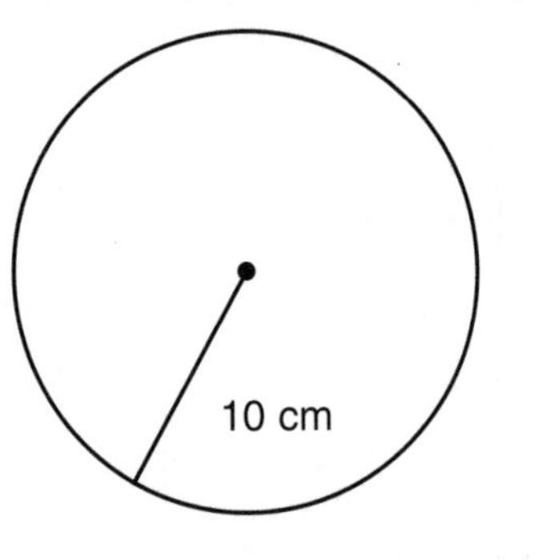

A = ____________________

5.

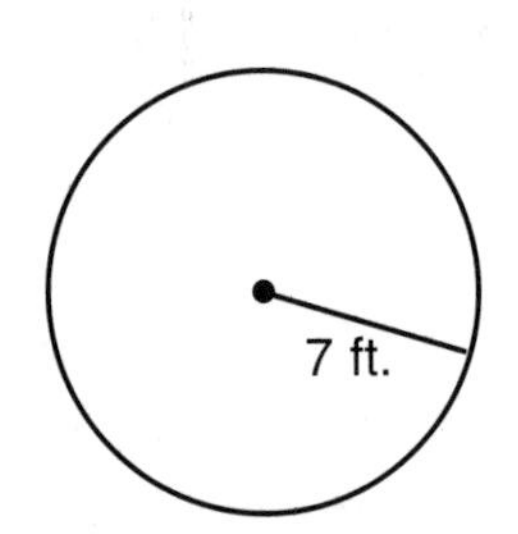

A = ____________________

6.

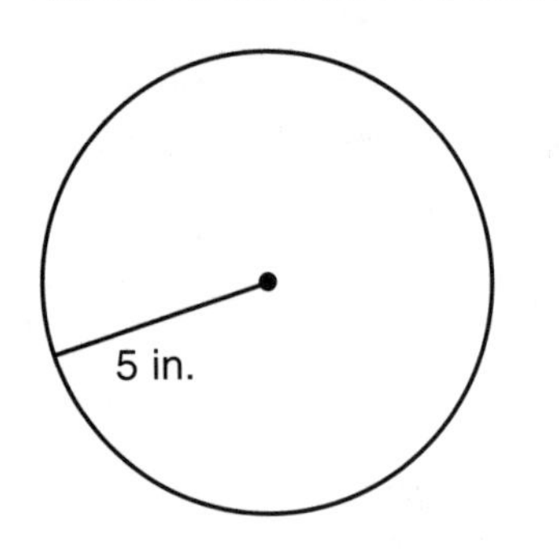

A = ____________________

7.

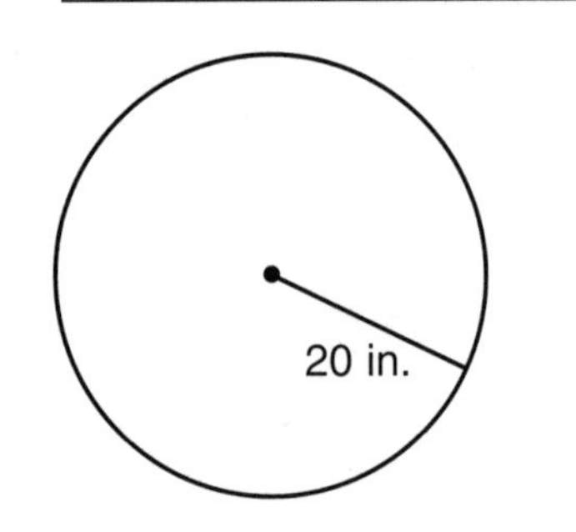

A = ____________________

8.

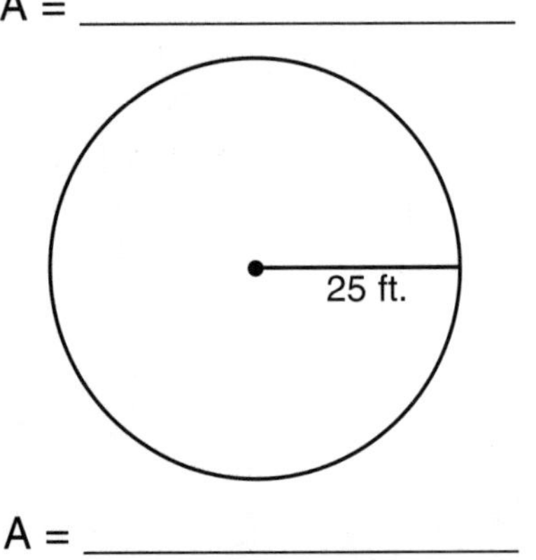

A = ____________________

Practice 30

Reminders

- A line of symmetry is a line drawn through the center of a flat shape so that one half of the shape can be folded to fit exactly over the other half.
- A figure may have one line of symmetry, several lines of symmetry, or no lines of symmetry.

Directions: Draw one line of symmetry through the symmetrical figures below. Circle the figures which have no lines of symmetry.

1. 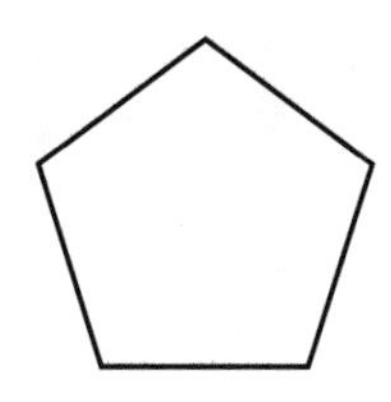

2. 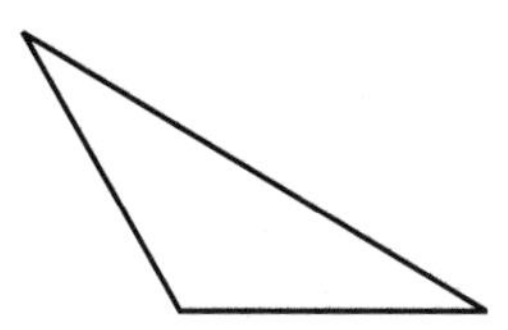

3. 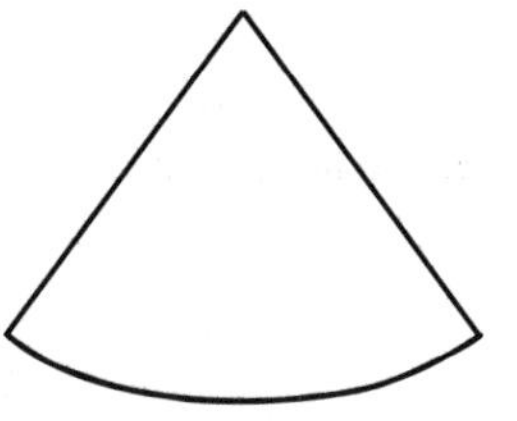

4. 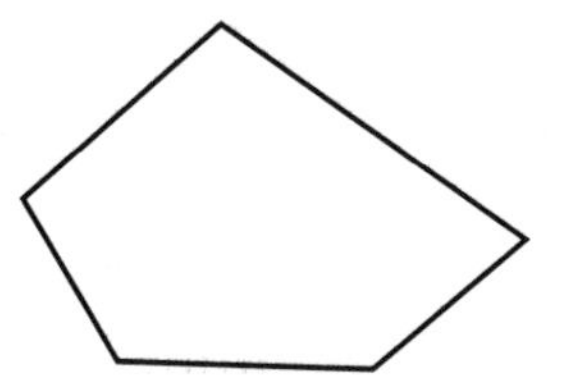

5. 

6. 

7. 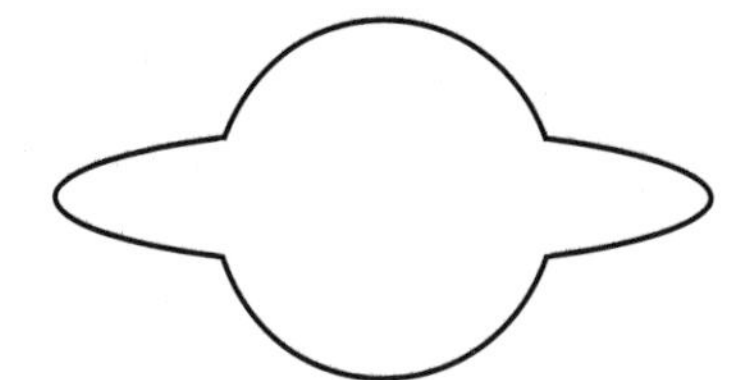

8. 

9. 

10. 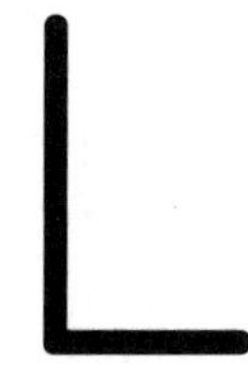

11. 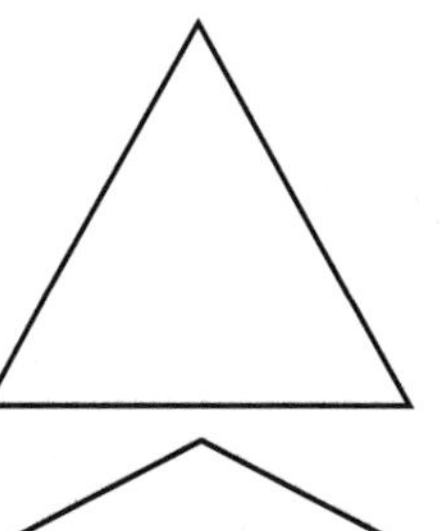

12. 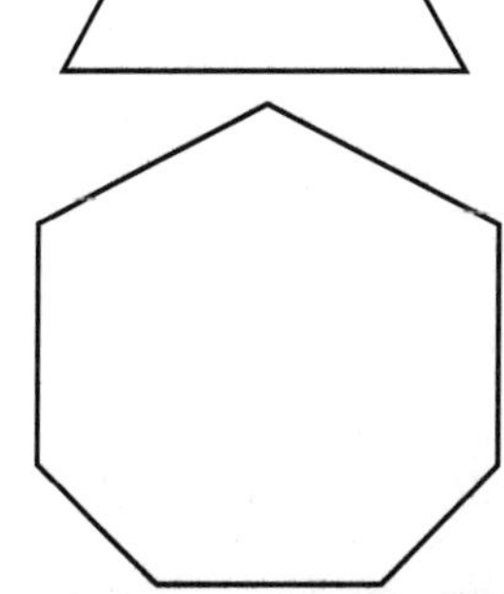

Practice 31

Reminders

- A line of symmetry is a line drawn through the center of a flat shape so that one half of the shape can be folded to fit exactly over the other half.
- A figure may have one line of symmetry, several lines of symmetry, or no lines of symmetry.

Directions: Draw two or more lines of symmetry through each figure below.

1. 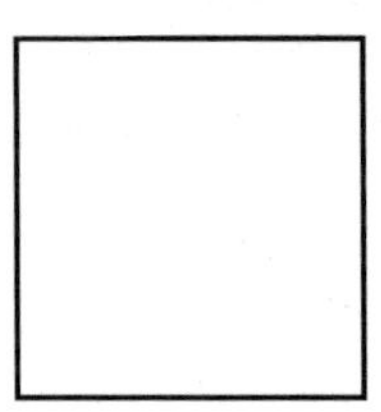

2. 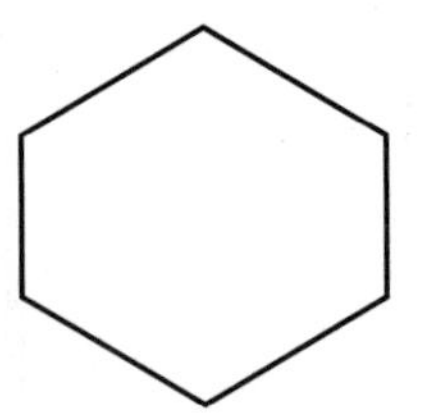

3. 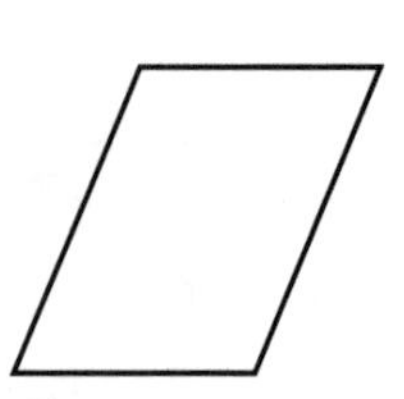

4. 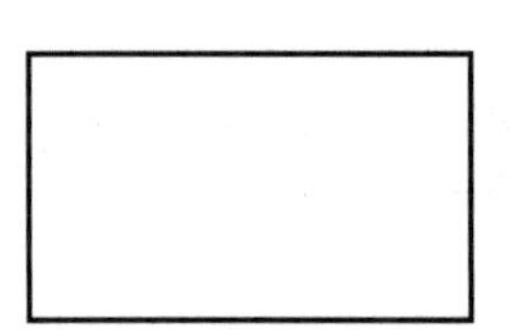

5. 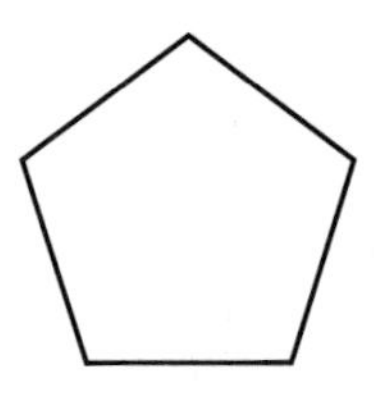

6. 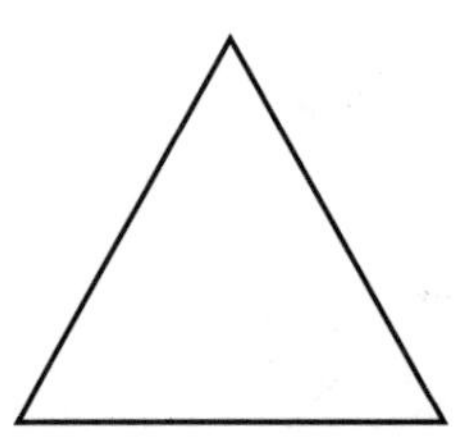

7. 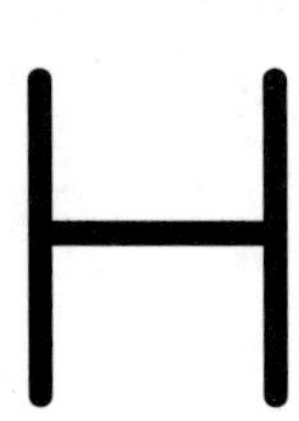

8. 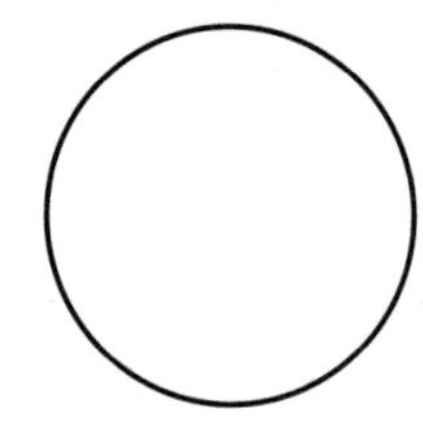

9. 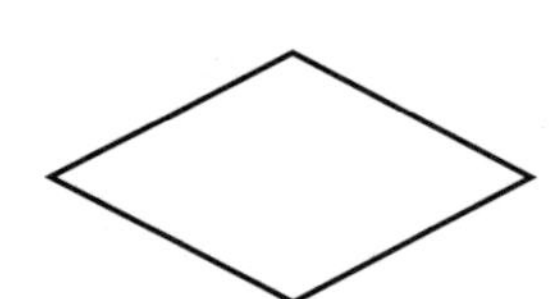

10. 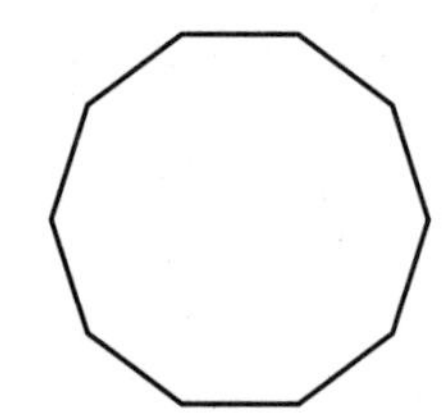

11. 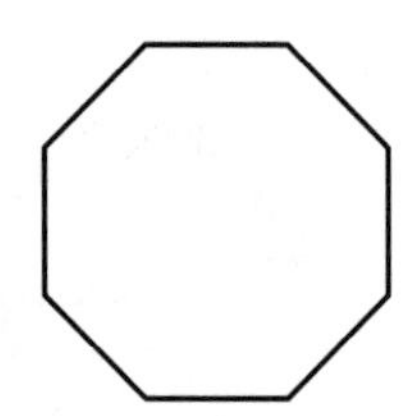

12. 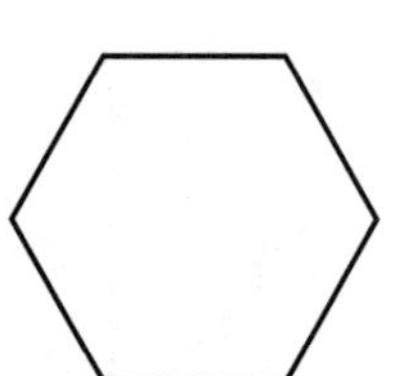

Practice 32

Reminders

- Congruent figures fit exactly over each other.
- Congruent figures are exactly the same in shape and size.
- Congruent figures can be turned over or around to fit.

Directions: Determine which of the figures in each set are congruent. Circle the congruent shapes.

1.

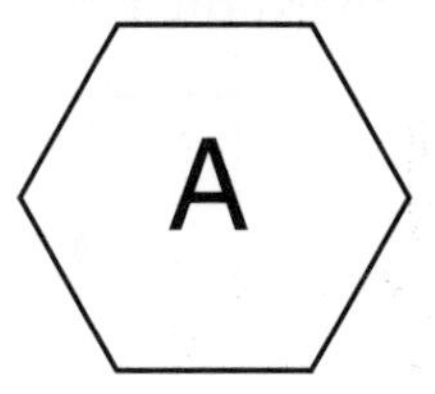

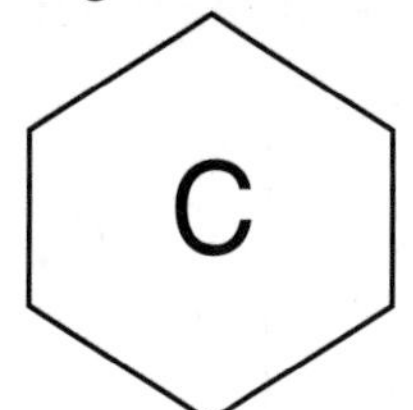

2.

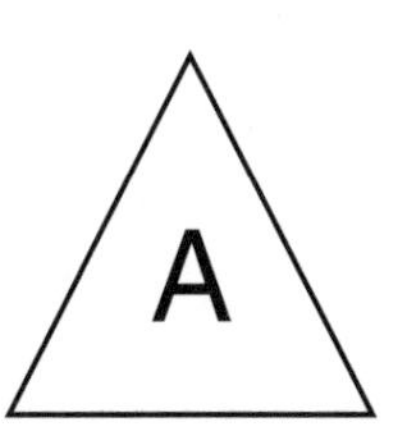

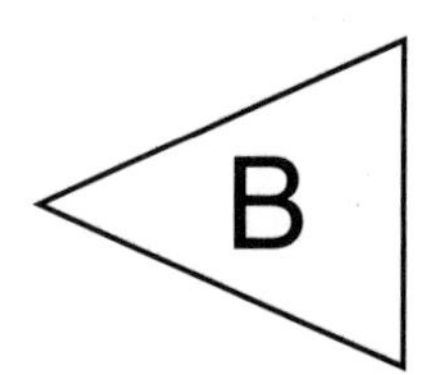

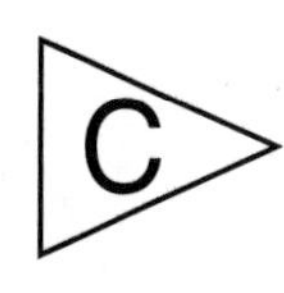

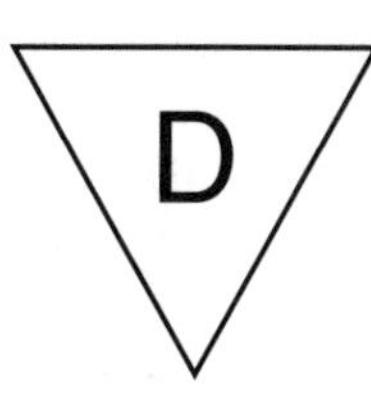

3.

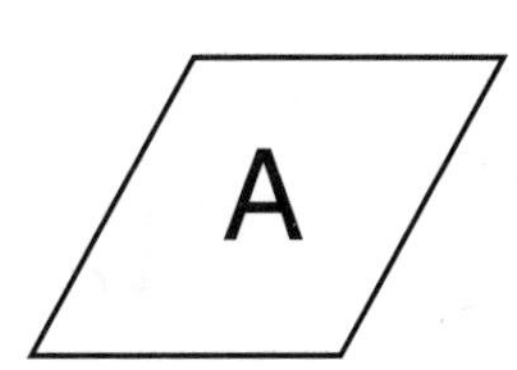

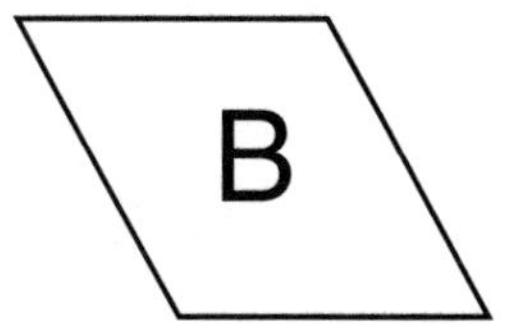

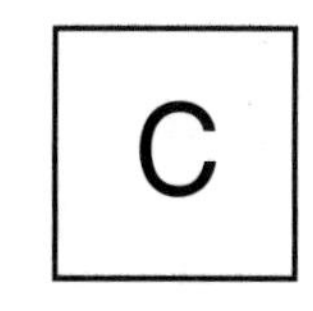

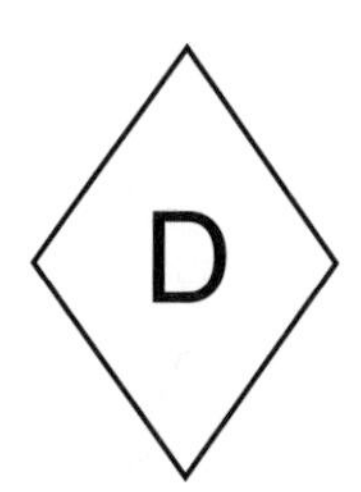

4.

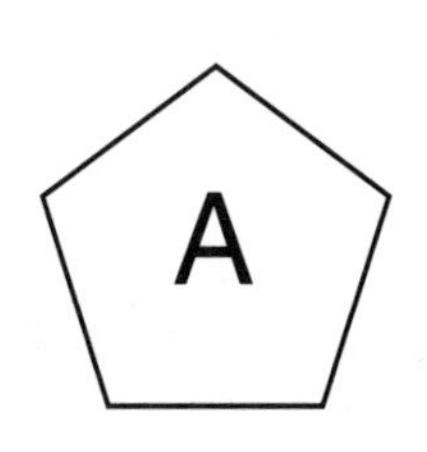

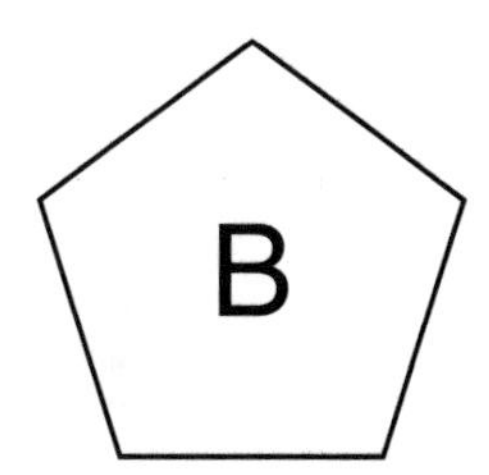

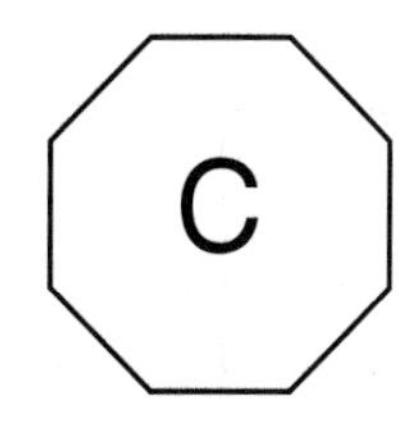

5.

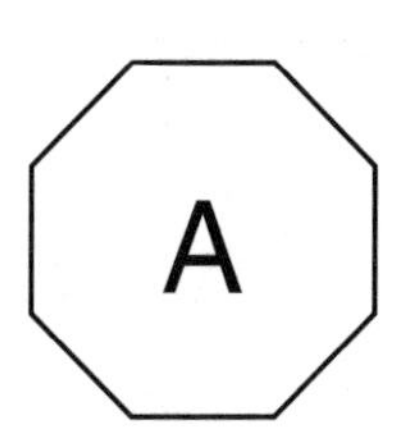

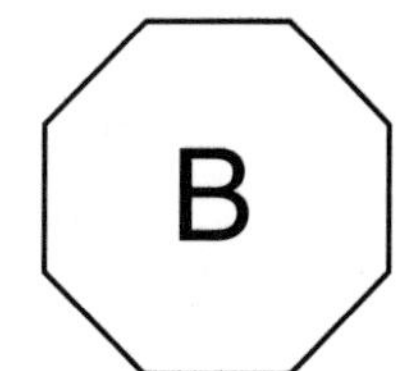

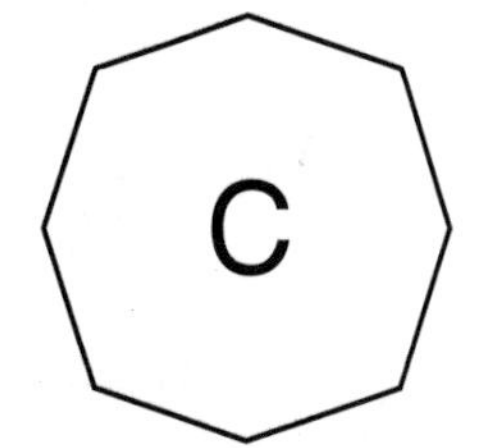

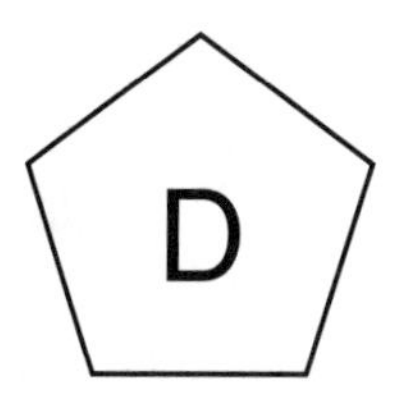

6.

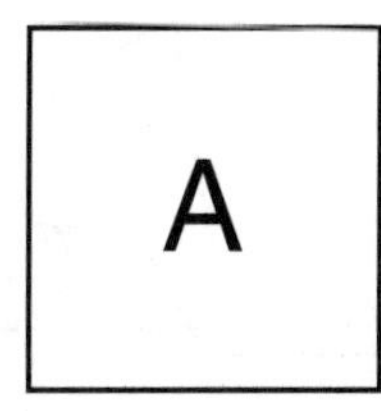

B

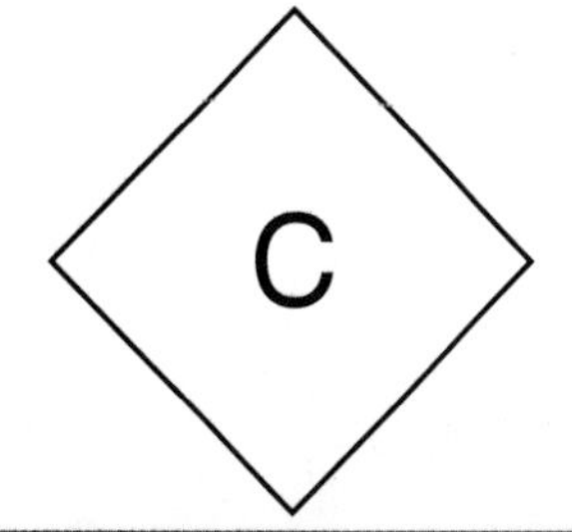

Practice 33

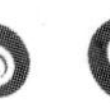

Reminders

- Similar figures are the same in shape but different in size.
- One similar figure is larger than the other.
- The corresponding angles in each similar figure will be equal.

Directions: Determine which of the figures in each set are similar. Circle the similar shapes.

1.

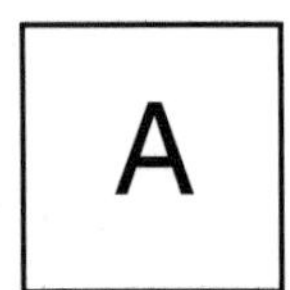

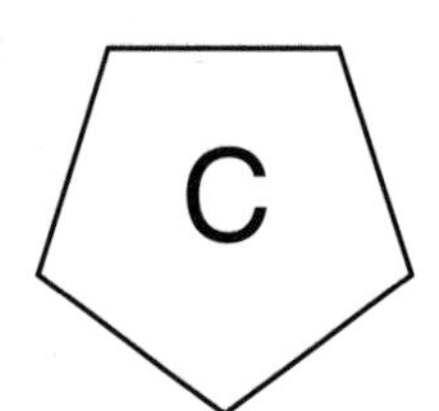

2.

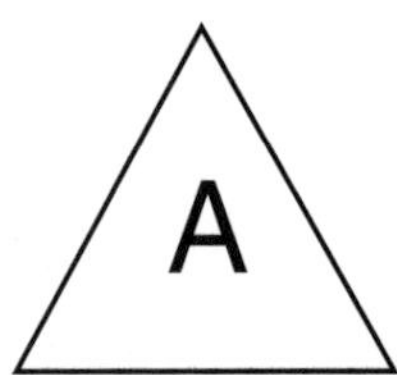

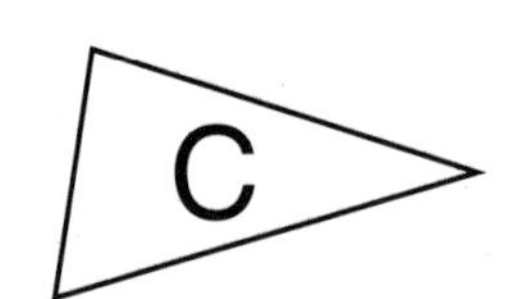

3.

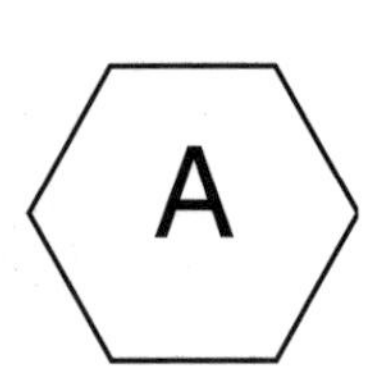

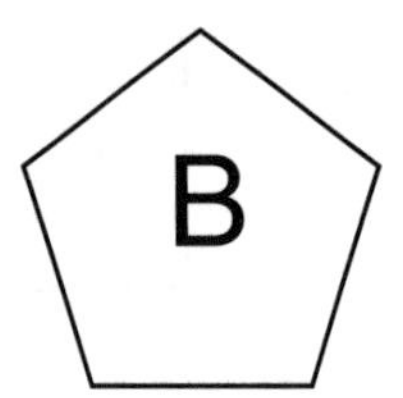

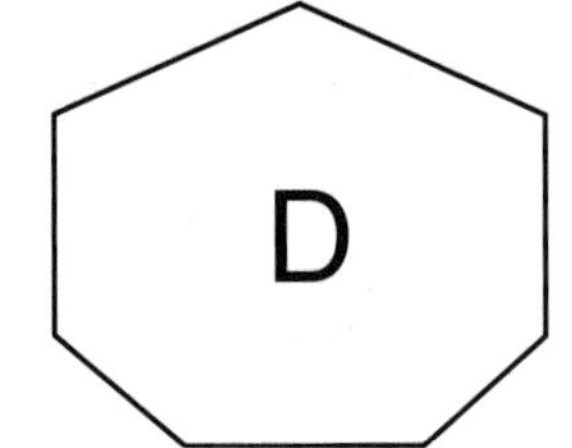

4.

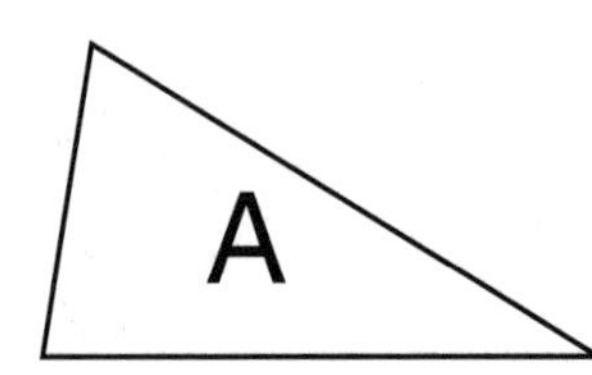

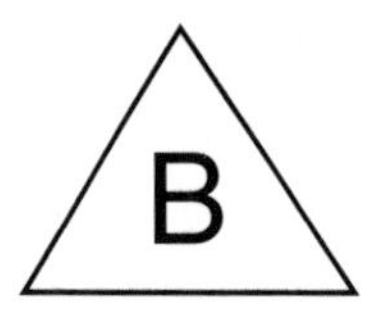

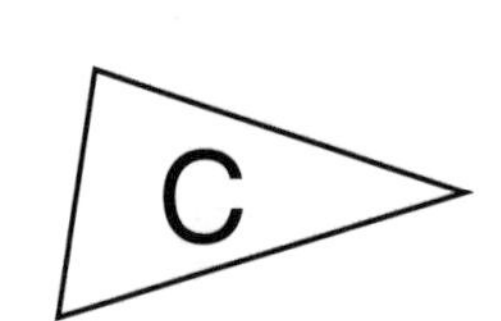

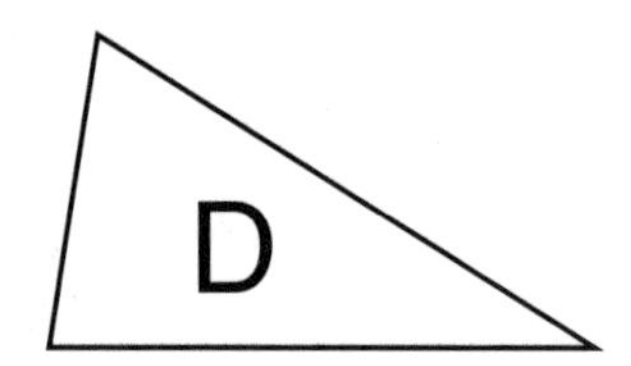

5.

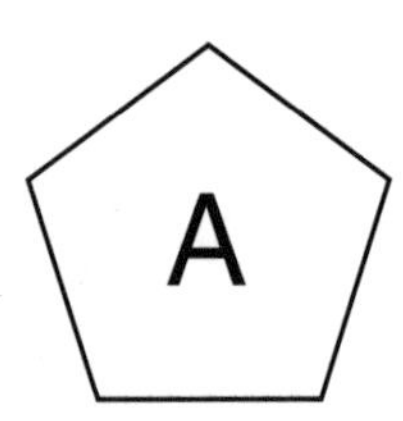

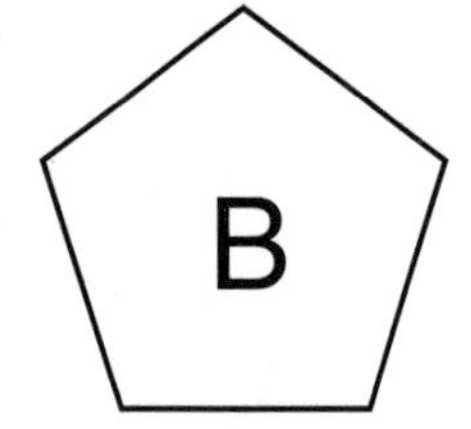

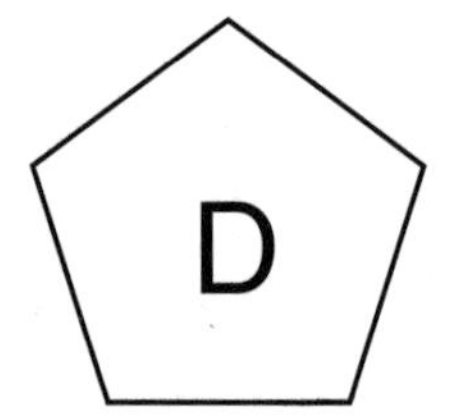

6.

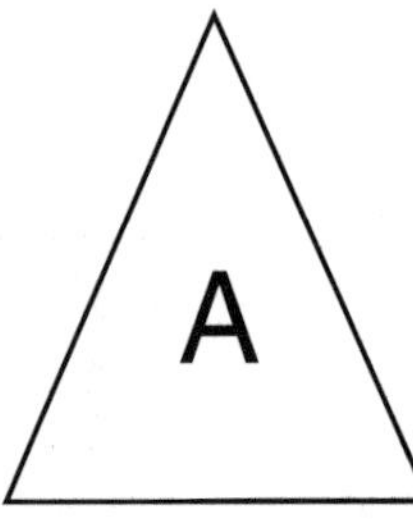

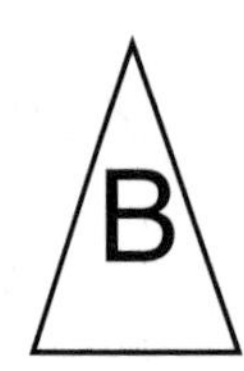

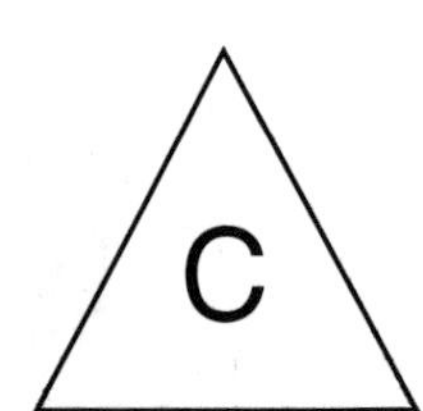

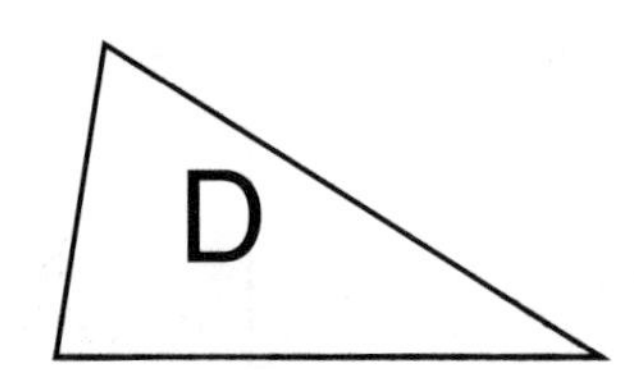

Practice 34

Reminders

- Congruent figures are exactly the same in shape and size.
- Congruent figures can be turned over or around to fit.
- Similar figures are the same in shape but different in size.

Directions: Determine which of the figures in each set are similar and which are congruent.

1. similar ____________ congruent____________

A B C D

2. similar ____________ congruent____________

A B C D

3. similar ____________ congruent____________

A B C D

4. similar ____________ congruent____________

A B C D

5. similar ____________ congruent____________

A B C D

6. similar ____________ congruent____________

A B C D

Practice 35

Directions: Identify each of these figures.

1. 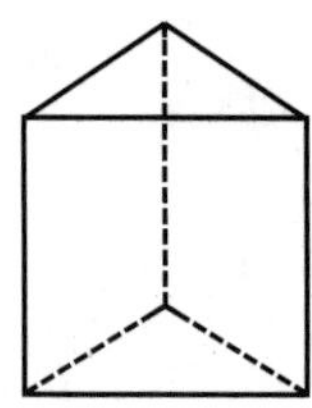

____

2. 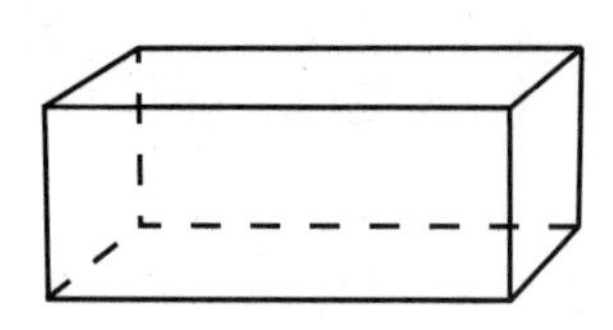

____

3. 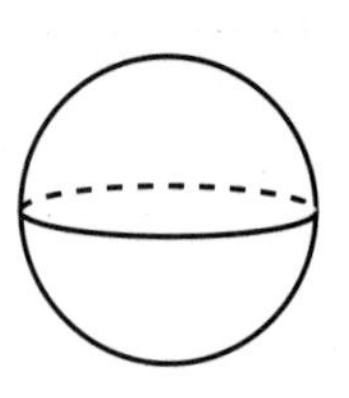

____

4. 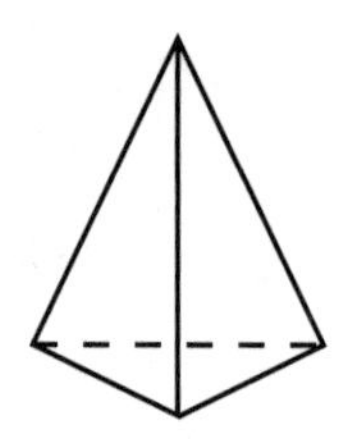

____

5. 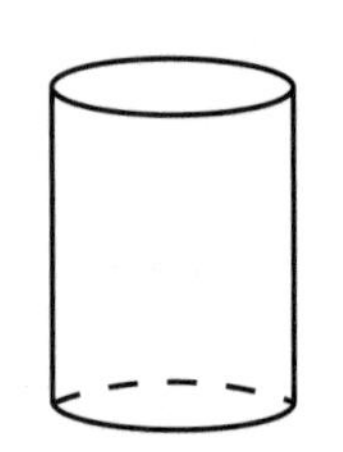

____

6. 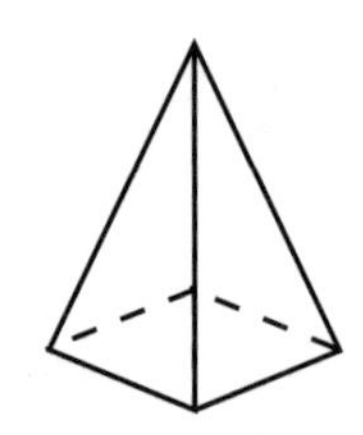

____

7. 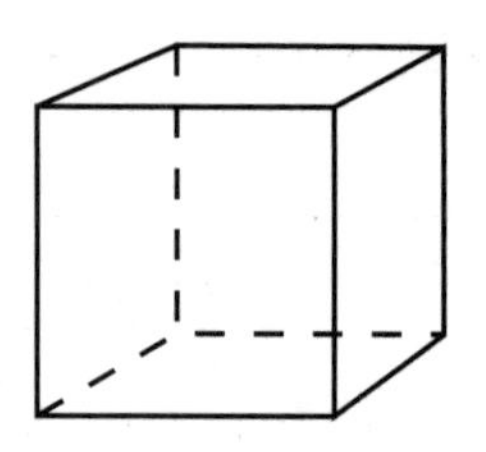

____

8. 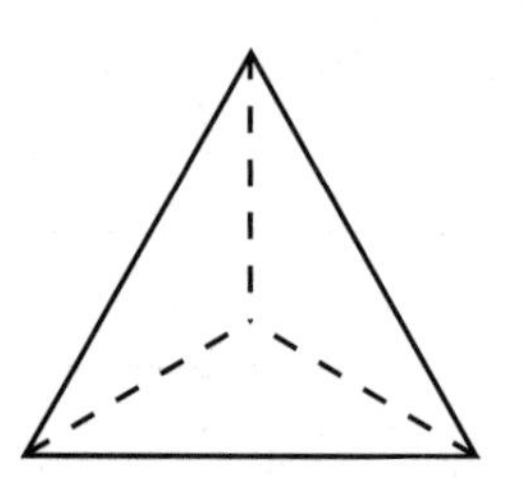

____

9. 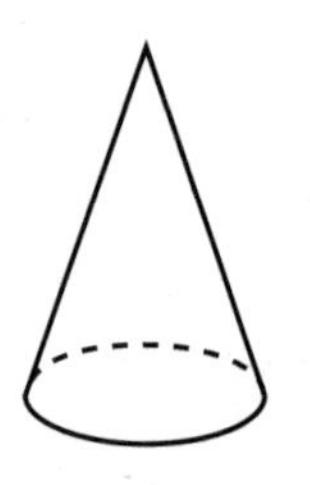

____

10. 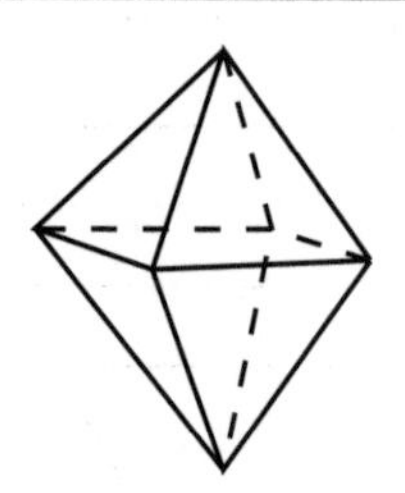

____

11. 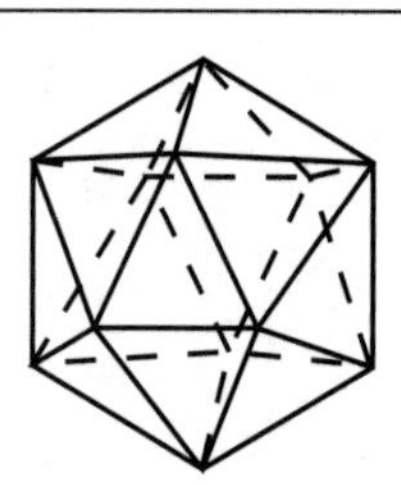

____

12. 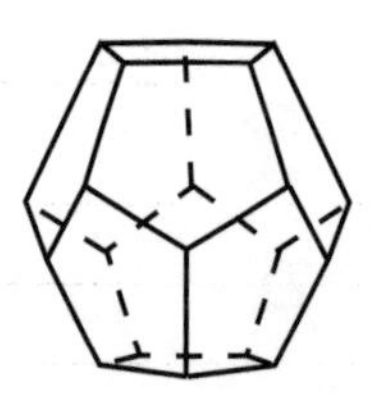

____

Practice 36

Reminders

- A face is the flat surface of a three dimensional figure.
- An edge is a line segment where two faces meet.
- A vertex is the point where edges meet.

Directions: Count the number of faces, edges, and vertices for each geometric solid on this page. Name each solid.

1.
name ______________________
faces ______________________
edges ______________________
vertices ______________________

2.
name ______________________
faces ______________________
edges ______________________
vertices ______________________

3.
name ______________________
faces ______________________
edges ______________________
vertices ______________________

4.
name ______________________
faces ______________________
edges ______________________
vertices ______________________

5.
name ______________________
faces ______________________
edges ______________________
vertices ______________________

6.
name ______________________
faces ______________________
edges ______________________
vertices ______________________

7.
name ______________________
faces ______________________
edges ______________________
vertices ______________________

8.
name ______________________
faces ______________________
edges ______________________
vertices ______________________

Test Practice 1

Directions: Identify each angle or shape. Use the most precise name. On the Answer Sheet, fill in the answer circle for your choice.

1.

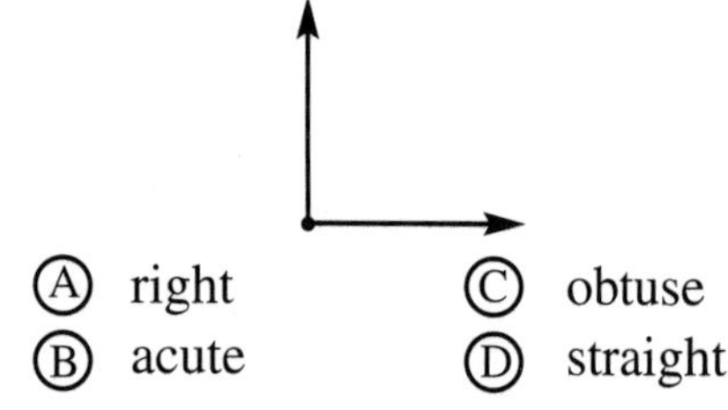

Ⓐ right Ⓒ obtuse
Ⓑ acute Ⓓ straight

2.

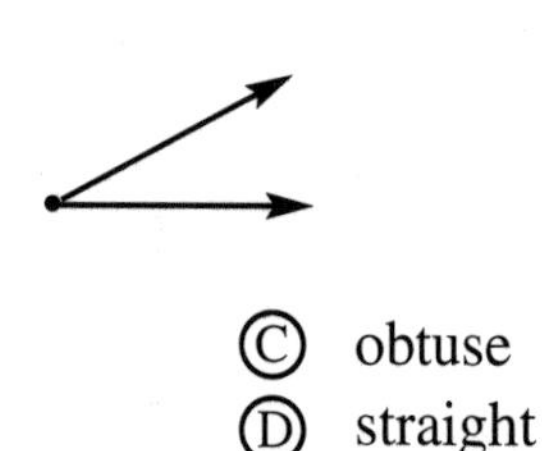

Ⓐ right Ⓒ obtuse
Ⓑ acute Ⓓ straight

3.

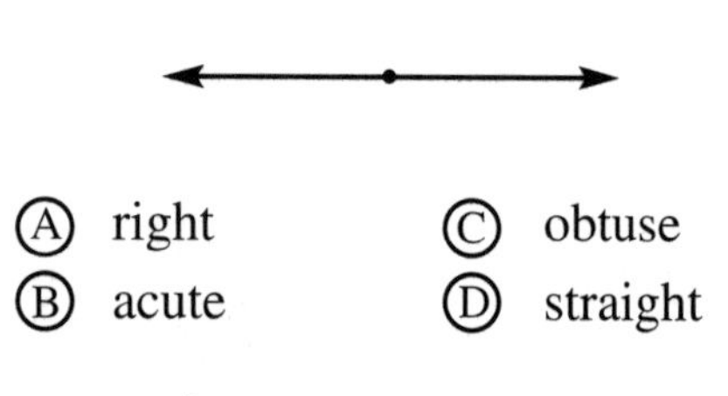

Ⓐ right Ⓒ obtuse
Ⓑ acute Ⓓ straight

4.

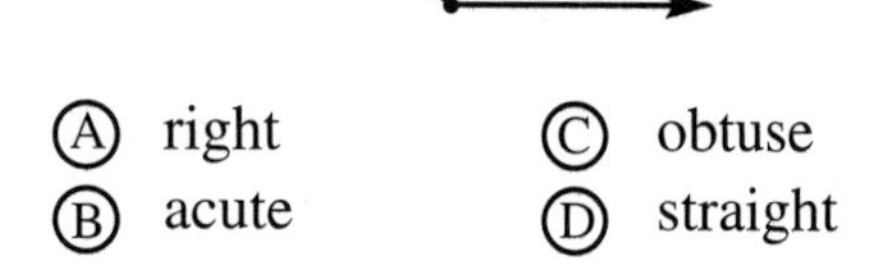

Ⓐ right Ⓒ obtuse
Ⓑ acute Ⓓ straight

5.

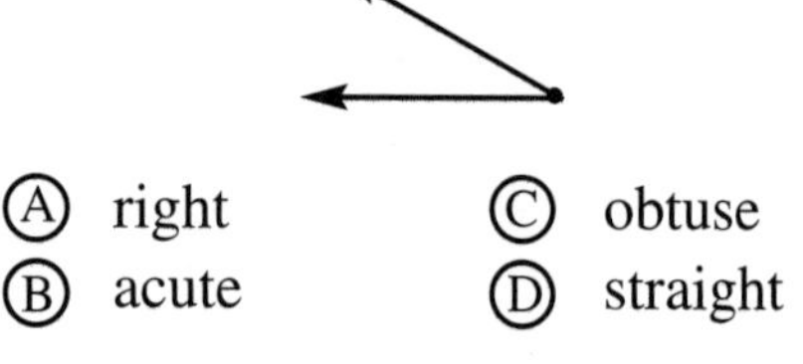

Ⓐ right Ⓒ obtuse
Ⓑ acute Ⓓ straight

6.

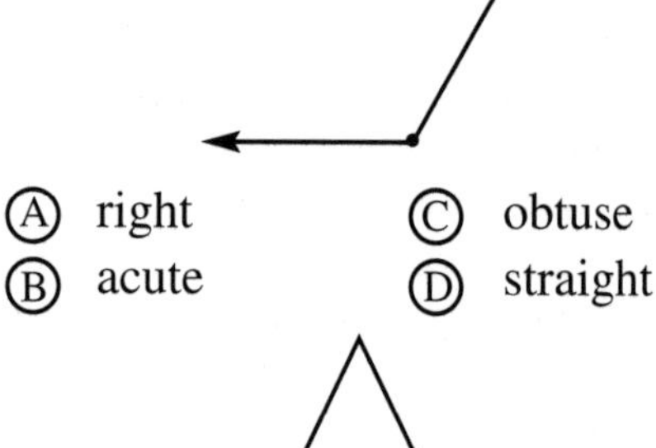

Ⓐ right Ⓒ obtuse
Ⓑ acute Ⓓ straight

7.

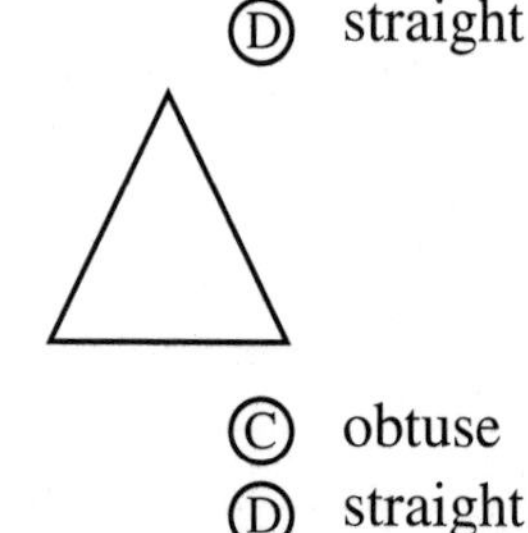

Ⓐ right Ⓒ obtuse
Ⓑ acute Ⓓ straight

8.

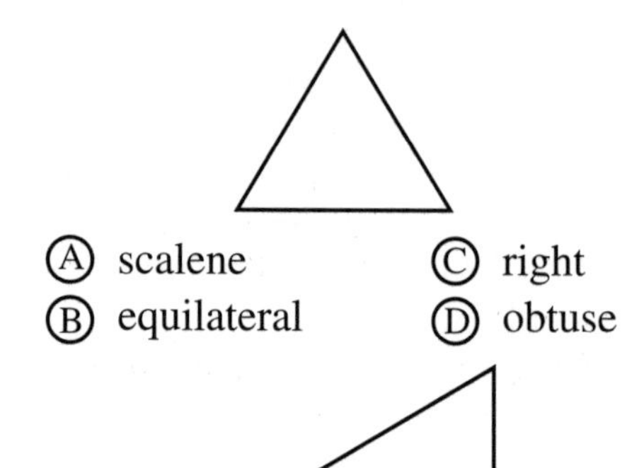

Ⓐ scalene Ⓒ right
Ⓑ equilateral Ⓓ obtuse

9.

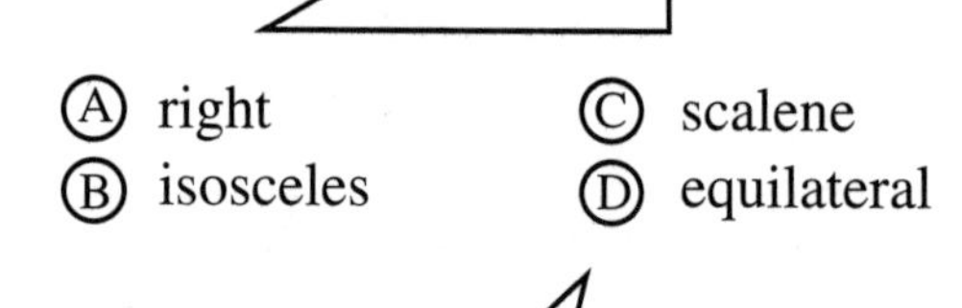

Ⓐ right Ⓒ scalene
Ⓑ isosceles Ⓓ equilateral

10.

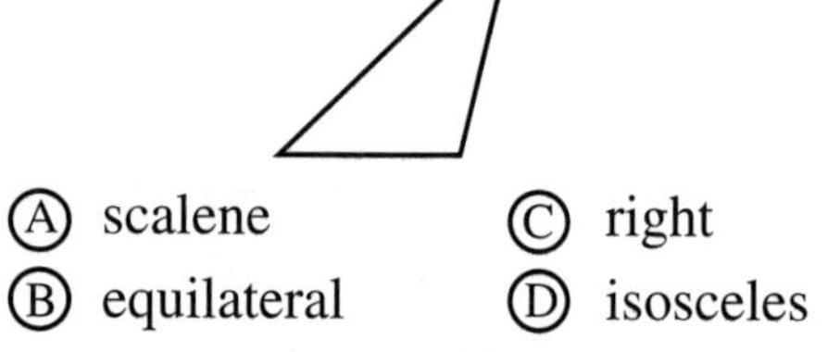

Ⓐ scalene Ⓒ right
Ⓑ equilateral Ⓓ isosceles

11.

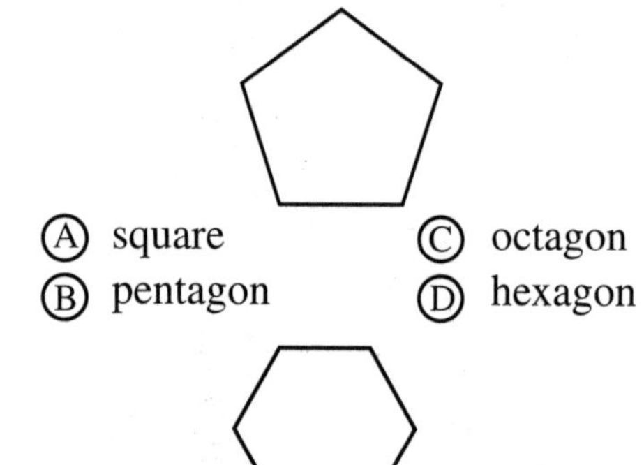

Ⓐ square Ⓒ octagon
Ⓑ pentagon Ⓓ hexagon

12.

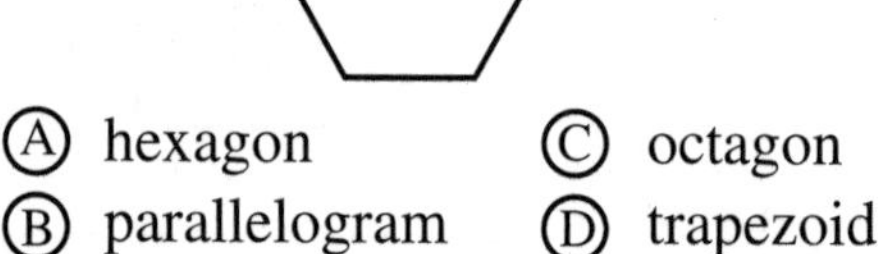

Ⓐ hexagon Ⓒ octagon
Ⓑ parallelogram Ⓓ trapezoid

13.

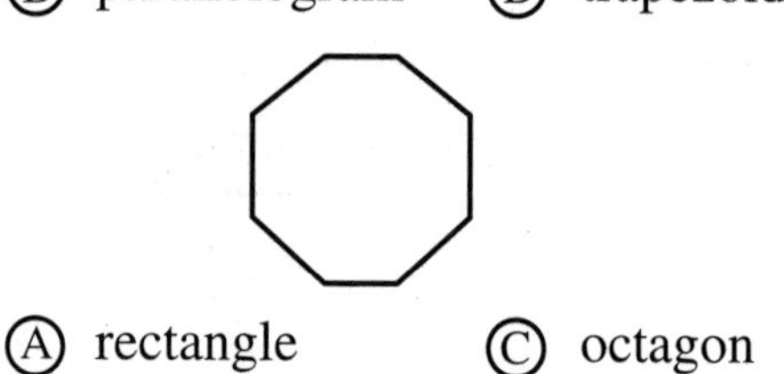

Ⓐ rectangle Ⓒ octagon
Ⓑ pentagon Ⓓ hexagon

14.

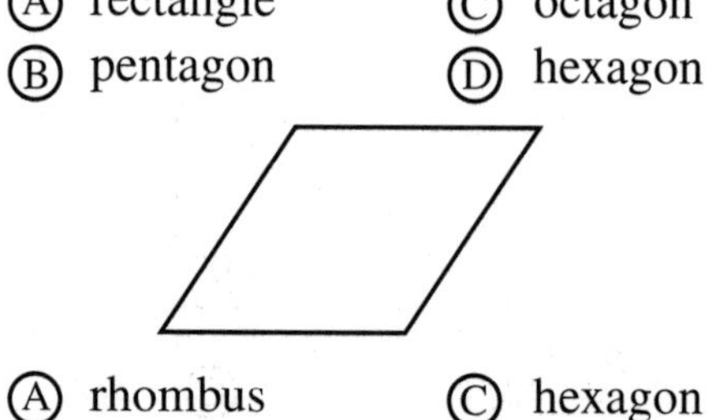

Ⓐ rhombus Ⓒ hexagon
Ⓑ pentagon Ⓓ trapezoid

Test Practice 2

Directions: Compute the missing angle in each figure. On the Answer Sheet, fill in the answer circle for your choice.

1.

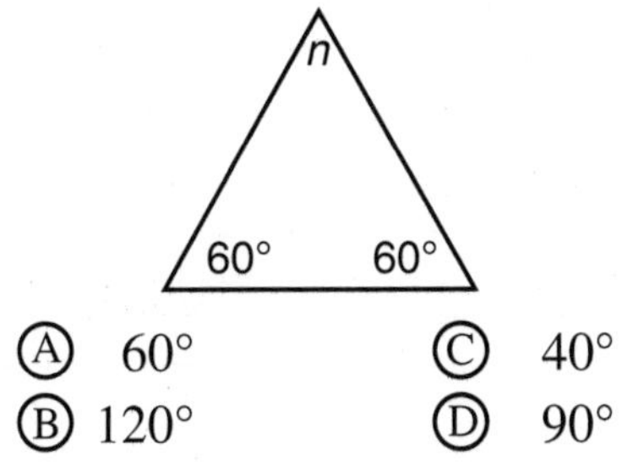

Ⓐ 60° Ⓒ 40°
Ⓑ 120° Ⓓ 90°

2.

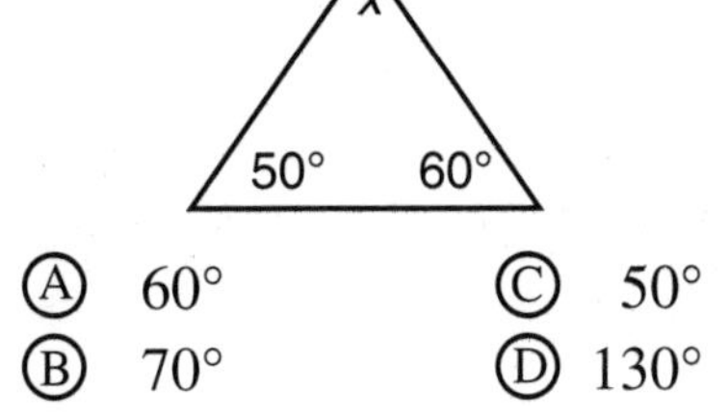

Ⓐ 60° Ⓒ 50°
Ⓑ 70° Ⓓ 130°

3.

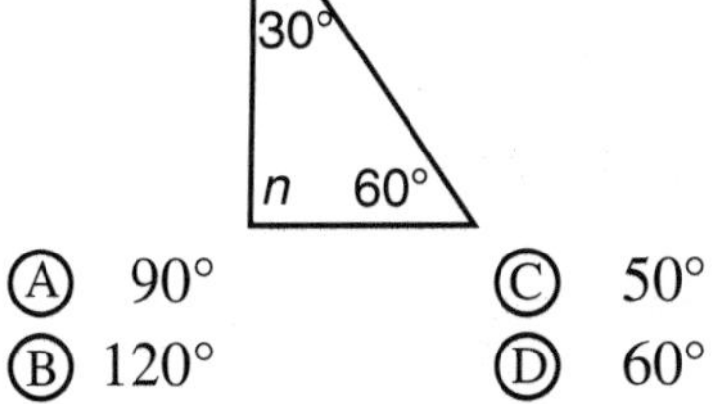

Ⓐ 90° Ⓒ 50°
Ⓑ 120° Ⓓ 60°

4.

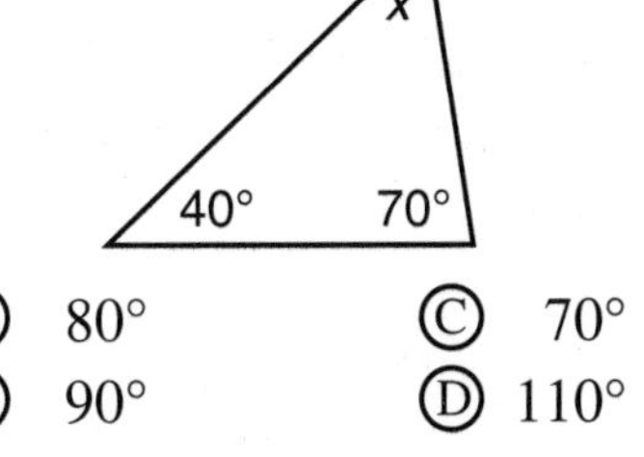

Ⓐ 80° Ⓒ 70°
Ⓑ 90° Ⓓ 110°

5.

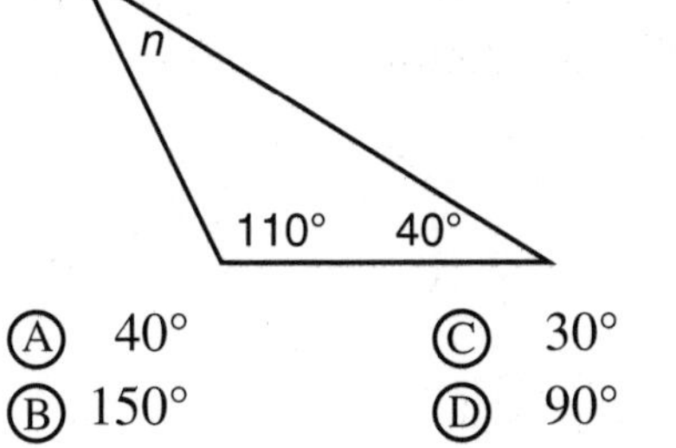

Ⓐ 40° Ⓒ 30°
Ⓑ 150° Ⓓ 90°

6.

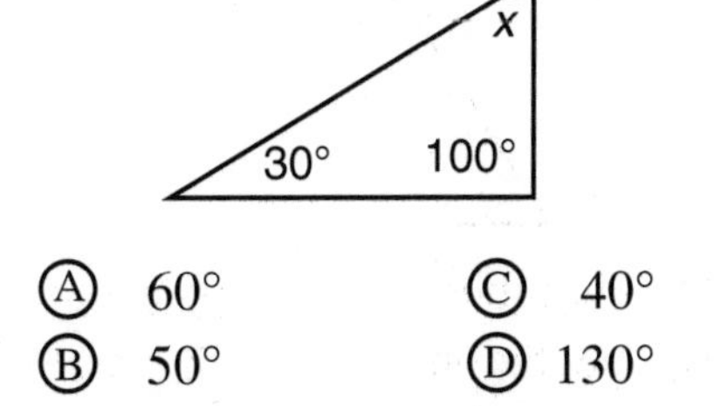

Ⓐ 60° Ⓒ 40°
Ⓑ 50° Ⓓ 130°

7.

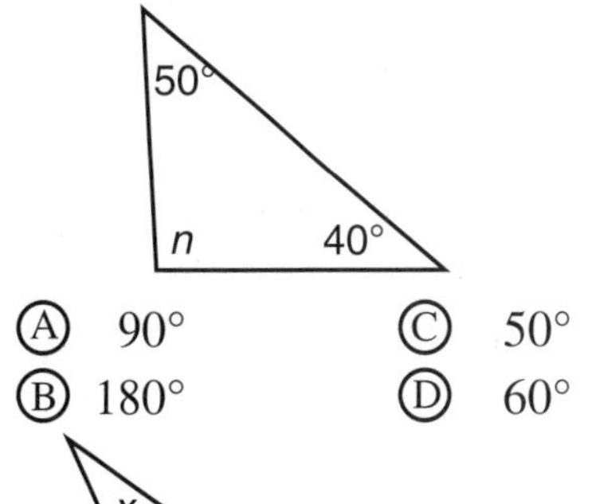

Ⓐ 90° Ⓒ 50°
Ⓑ 180° Ⓓ 60°

8.

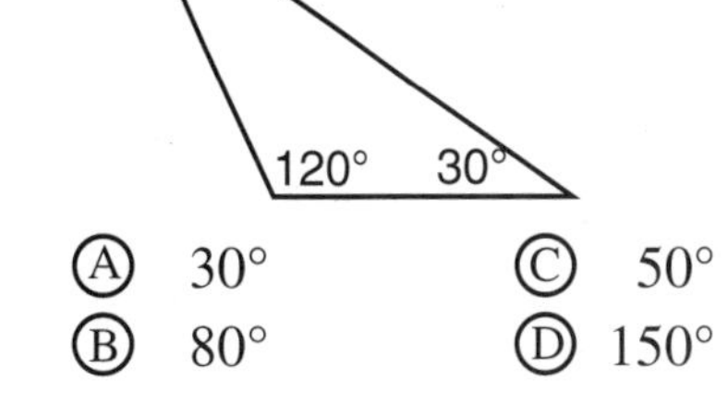

Ⓐ 30° Ⓒ 50°
Ⓑ 80° Ⓓ 150°

9.

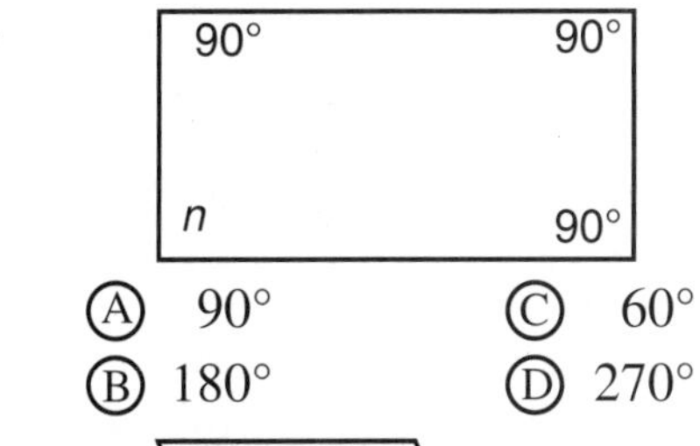

Ⓐ 90° Ⓒ 60°
Ⓑ 180° Ⓓ 270°

10.

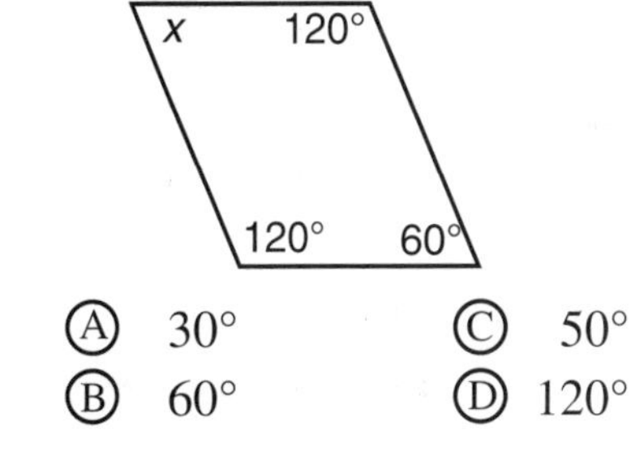

Ⓐ 30° Ⓒ 50°
Ⓑ 60° Ⓓ 120°

11.

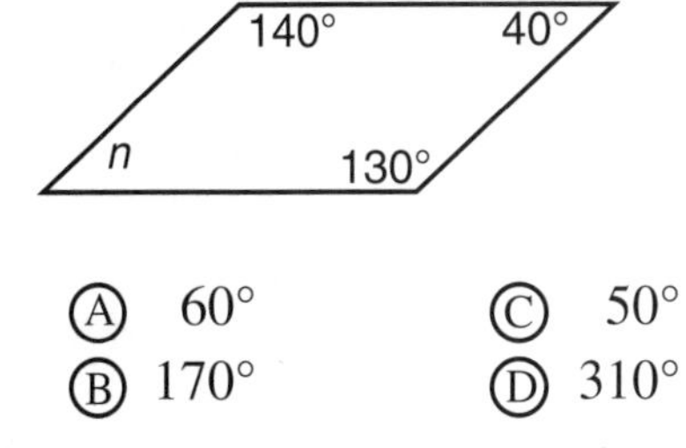

Ⓐ 60° Ⓒ 50°
Ⓑ 170° Ⓓ 310°

12.

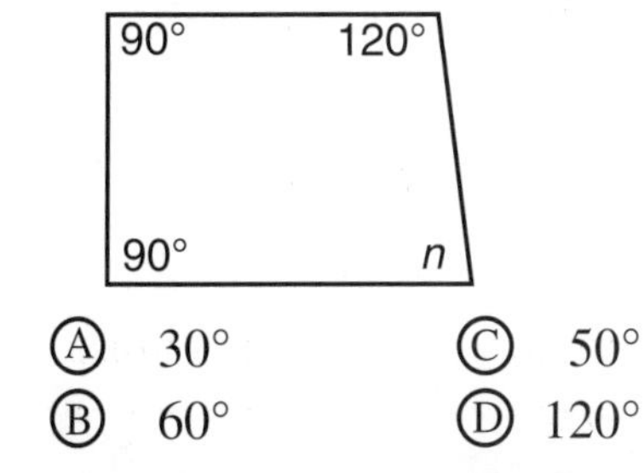

Ⓐ 30° Ⓒ 50°
Ⓑ 60° Ⓓ 120°

Test Practice 3

Directions: Answer each question. On the Answer Sheet, fill in the answer circle for your choice.

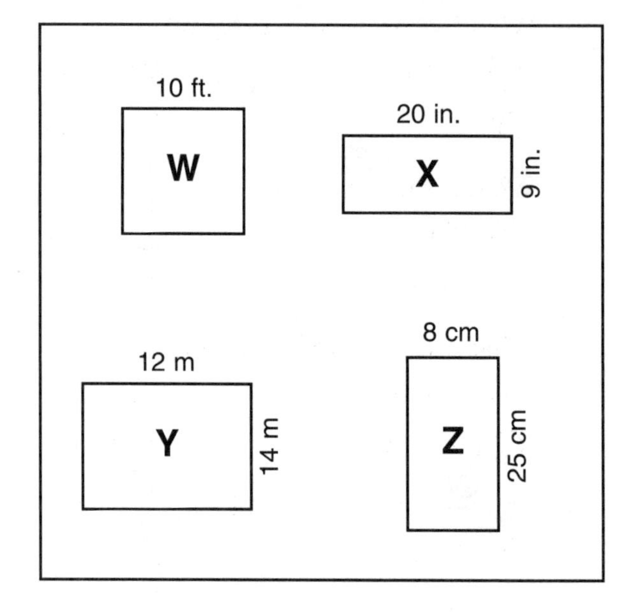

1. What is the perimeter of square W?

Ⓐ 40 ft. Ⓒ 20 ft.

Ⓑ 400 ft. Ⓓ 10 ft.

2. What is the perimeter of rectangle X?

Ⓐ 28 in. Ⓒ 40 in.

Ⓑ 180 in. Ⓓ 58 in.

3. What is the perimeter of rectangle Y?

Ⓐ 26 m Ⓒ 52 m

Ⓑ 168 m Ⓓ 24 m

4. What is the perimeter of rectangle Z?

Ⓐ 33 cm Ⓒ 56 cm

Ⓑ 66 cm Ⓓ 200 cm

5. What is the perimeter of a square with 25 ft. long sides?

Ⓐ 40 ft. Ⓒ 50 ft.

Ⓑ 100 ft. Ⓓ 625 ft.

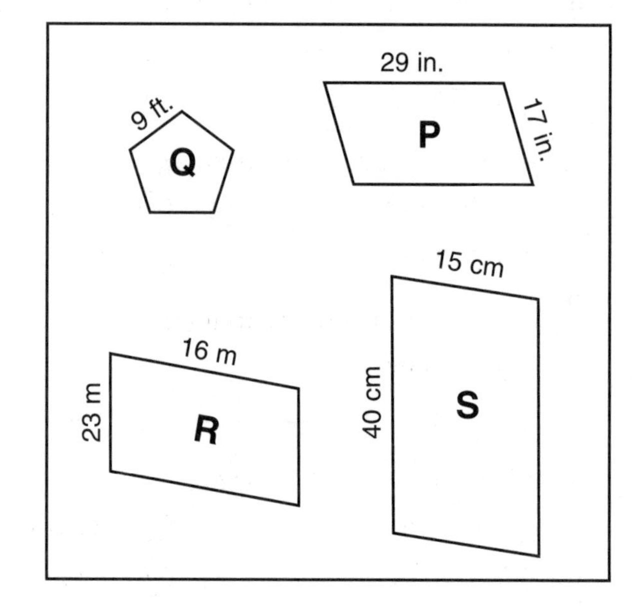

6. What is the perimeter of pentagon Q?

Ⓐ 45 ft. Ⓒ 36 ft.

Ⓑ 54 ft. Ⓓ 63 ft.

7. What is the perimeter of parallelogram P?

Ⓐ 56in. Ⓒ 92 in.

Ⓑ 46 in. Ⓓ 493 in.

8. What is the perimeter of parallelogram R?

Ⓐ 78 m Ⓒ 68 m

Ⓑ 39 m Ⓓ 368 m

9. What is the perimeter of parallelogram S?

Ⓐ 105 cm Ⓒ 110 cm

Ⓑ 55 cm Ⓓ 2,200 cm

10. What is the perimeter of an octagon with 15 ft. long sides?

Ⓐ 220 ft. Ⓒ 120 ft.

Ⓑ 90 ft. Ⓓ 75 ft.

Test Practice 4

Directions: Answer each question. On the Answer Sheet, fill in the answer circle for your choice.

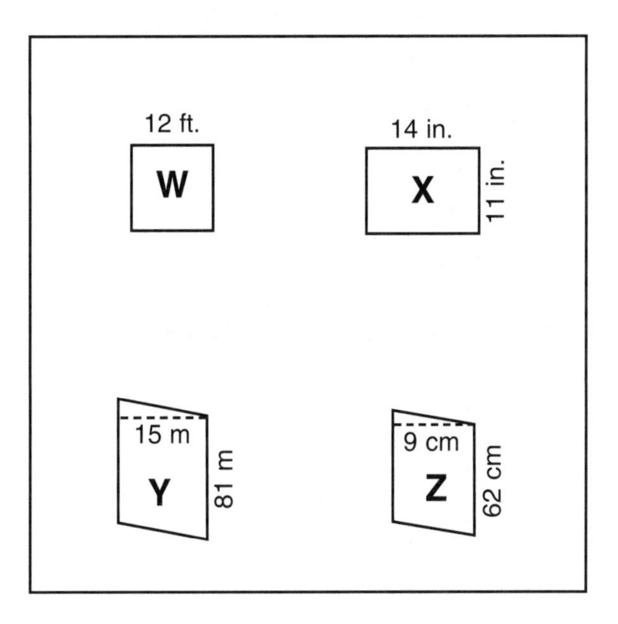

1. What is the area of square W?
   - Ⓐ 24 ft.^{2}
   - Ⓑ 144 ft.^{2}
   - Ⓒ 48 ft.^{2}
   - Ⓓ 244 ft.^{2}

2. What is the area of rectangle X?
   - Ⓐ 154 in.^{2}
   - Ⓑ 140 in.^{2}
   - Ⓒ 25 in.^{2}
   - Ⓓ 50 in.^{2}

3. What if the area of parallelogram Y?
   - Ⓐ 96 m^2
   - Ⓑ 1,215 m^2
   - Ⓒ 188 m^2
   - Ⓓ 1,315 m^2

4. What is the area of parallelogram Z?
   - Ⓐ 96 cm^2
   - Ⓑ 141 cm^2
   - Ⓒ 558 cm^2
   - Ⓓ 578 cm^2

5. What is the area of a square with 15 ft. long sides?
   - Ⓐ 60 ft.^{2}
   - Ⓑ 225 ft.^{2}
   - Ⓒ 30 ft.^{2}
   - Ⓓ 625 ft.^{2}

---

6. What is the area of the triangle Q?
   - Ⓐ 63 ft.^{2}
   - Ⓑ 126 ft.^{2}
   - Ⓒ 50 ft.^{2}
   - Ⓓ 136 ft.^{2}

7. What is the area of the triangle P?
   - Ⓐ 104 in.^{2}
   - Ⓑ 240 in.^{2}
   - Ⓒ 52 in.^{2}
   - Ⓓ 480 in.^{2}

8. What is the area of the triangle R?
   - Ⓐ 78 m^2
   - Ⓑ 338 m^2
   - Ⓒ 88 m^2
   - Ⓓ 169 m^2

9. What is the area of the triangle S?
   - Ⓐ 448 cm^2
   - Ⓑ 548 cm^2
   - Ⓒ 224 cm^2
   - Ⓓ 88 cm^2

10. What is the area of the triangle with a 10 foot base and a 32 ft. height?
    - Ⓐ 160 ft.^{2}
    - Ⓑ 320 ft.^{2}
    - Ⓒ 42 ft.^{2}
    - Ⓓ 180 ft.^{2}

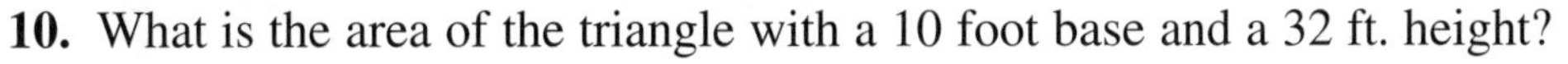

Test Practice 5

Directions: Answer each question. On the Answer Sheet, fill in the answer circle for your choice.

10 ft.
X
12 in.
Y
4 in.
6 in.
20 m
Z
12 m
8 m

1. What is the volume of cube X?

Ⓐ 30 ft.^{3} Ⓒ 1,000 ft.^{3}
Ⓑ 100 ft.^{3} Ⓓ 110 ft.^{3}

2. What is the volume of rectangular prism Y?

Ⓐ 20 in.^{3} Ⓒ 480 in.^{3}
Ⓑ 288 in.^{3} Ⓓ 48 in.^{3}

3. What is the volume of rectangular prism Z?

Ⓐ 1,920 m^3 Ⓒ 40m^3
Ⓑ 192 m^3 Ⓓ 160 m^3

4. What is the volume of a cube 4 feet on each edge?

Ⓐ 12 ft.^{3} Ⓒ 64 ft.^{3}
Ⓑ 16 ft.^{3} Ⓓ 256 ft.^{3}

5. What is the volume of a rectangular prism 3 yards long, 4 yards wide, and 5 yards high?

Ⓐ 12 yd.^{3} Ⓒ 17 yd.^{3}
Ⓑ 60 yd.^{3} Ⓓ 600 yd.^{3}

---

6. What is the circumference of circle P?

Ⓐ 12.56 ft. Ⓒ 1,256 ft.
Ⓑ 7.14 ft. Ⓓ 12 ft.

7. What is the circumference of circle Q?

Ⓐ 30 ft. Ⓒ 314 ft.
Ⓑ 31.4 ft. Ⓓ 34 ft.

8. What is the circumference of circle R?

Ⓐ 25.12 m Ⓒ 11.12 m
Ⓑ 251.2 m Ⓓ 24 m

9. What is the area of circle P?

Ⓐ 12.56 ft.^{2} Ⓒ 100 ft.^{2}
Ⓑ 50.12 ft.^{2} Ⓓ 19.14 ft.^{2}

10. What is the area of circle Q?

Ⓐ 7.85 in.^{2} Ⓒ 314 in.^{2}
Ⓑ 78.5 in.^{2} Ⓓ 3140 in.^{2}

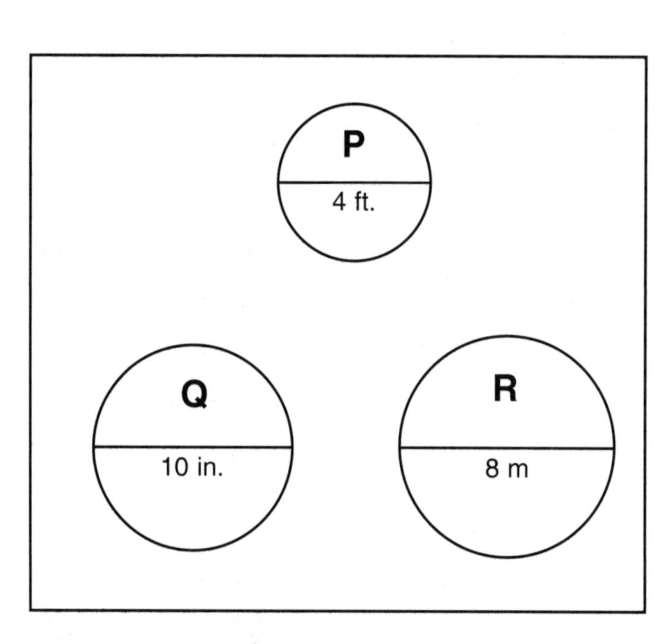

Remember: Pi = 3.14 $C = \pi d$ $A = \pi r^2$

Test Practice 6

Directions: Answer each question. On the Answer Sheet, fill in the answer circle for your choice.

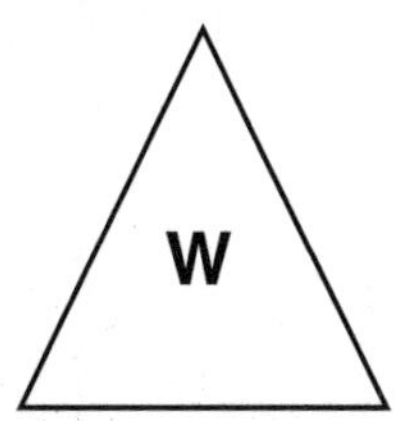

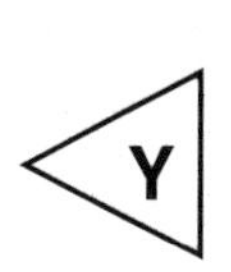

1. Which figures are congruent?
Ⓐ W and X Ⓑ W and Y Ⓒ Y and Z Ⓓ X and Y

2. Which figures are similar?
Ⓐ W and Y Ⓑ X and Y Ⓒ W and Z Ⓓ X and Z

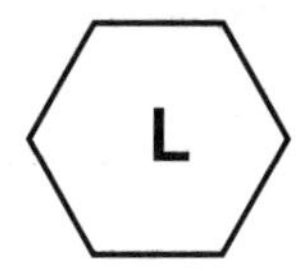

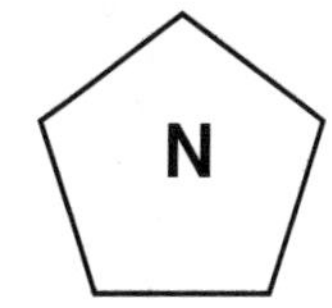

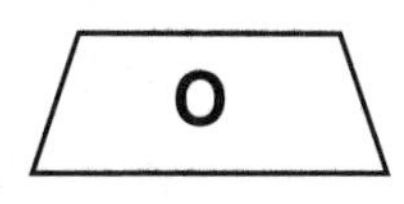

3. Which figure is a pentagon?
Ⓐ M Ⓑ L Ⓒ N Ⓓ O

4. Which figure is an octagon?
Ⓐ M Ⓑ L Ⓒ N Ⓓ O

5. Which figure is a hexagon?
Ⓐ M Ⓑ L Ⓒ N Ⓓ O

6. Which figure is a trapezoid?
Ⓐ M Ⓑ L Ⓒ O Ⓓ N

7. On which figure can you draw five lines of symmetry?
Ⓐ M Ⓑ N Ⓒ L Ⓓ O

8. On which figure can you draw only one line of symmetry?
Ⓐ L Ⓑ M Ⓒ N Ⓓ O

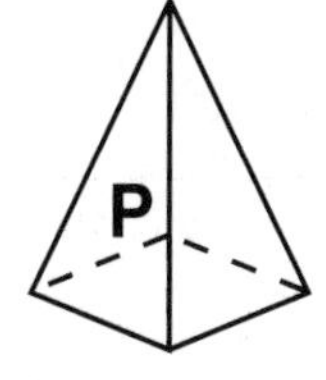

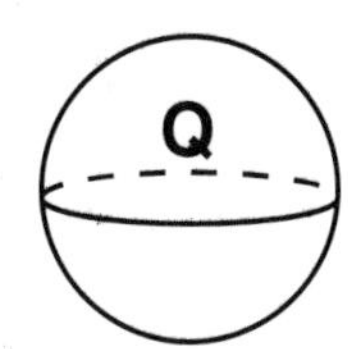

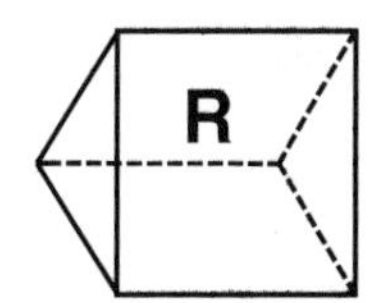

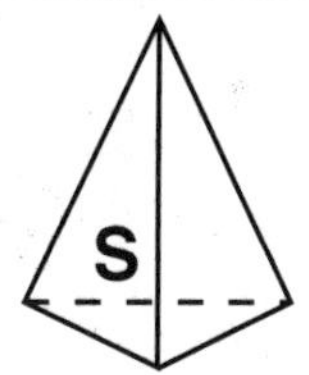

9. Which figure is a square pyramid?
Ⓐ P Ⓑ Q Ⓒ R Ⓓ S

10. Which figure is a triangular prism?
Ⓐ P Ⓑ Q Ⓒ R Ⓓ S

Answer Sheet

Test Practice 1	Test Practice 2	Test Practice 3
1. Ⓐ Ⓑ Ⓒ Ⓓ	1. Ⓐ Ⓑ Ⓒ Ⓓ	1. Ⓐ Ⓑ Ⓒ Ⓓ
2. Ⓐ Ⓑ Ⓒ Ⓓ	2. Ⓐ Ⓑ Ⓒ Ⓓ	2. Ⓐ Ⓑ Ⓒ Ⓓ
3. Ⓐ Ⓑ Ⓒ Ⓓ	3. Ⓐ Ⓑ Ⓒ Ⓓ	3. Ⓐ Ⓑ Ⓒ Ⓓ
4. Ⓐ Ⓑ Ⓒ Ⓓ	4. Ⓐ Ⓑ Ⓒ Ⓓ	4. Ⓐ Ⓑ Ⓒ Ⓓ
5. Ⓐ Ⓑ Ⓒ Ⓓ	5. Ⓐ Ⓑ Ⓒ Ⓓ	5. Ⓐ Ⓑ Ⓒ Ⓓ
6. Ⓐ Ⓑ Ⓒ Ⓓ	6. Ⓐ Ⓑ Ⓒ Ⓓ	6. Ⓐ Ⓑ Ⓒ Ⓓ
7. Ⓐ Ⓑ Ⓒ Ⓓ	7. Ⓐ Ⓑ Ⓒ Ⓓ	7. Ⓐ Ⓑ Ⓒ Ⓓ
8. Ⓐ Ⓑ Ⓒ Ⓓ	8. Ⓐ Ⓑ Ⓒ Ⓓ	8. Ⓐ Ⓑ Ⓒ Ⓓ
9. Ⓐ Ⓑ Ⓒ Ⓓ	9. Ⓐ Ⓑ Ⓒ Ⓓ	9. Ⓐ Ⓑ Ⓒ Ⓓ
10. Ⓐ Ⓑ Ⓒ Ⓓ	10. Ⓐ Ⓑ Ⓒ Ⓓ	10. Ⓐ Ⓑ Ⓒ Ⓓ
11. Ⓐ Ⓑ Ⓒ Ⓓ	11. Ⓐ Ⓑ Ⓒ Ⓓ	
12. Ⓐ Ⓑ Ⓒ Ⓓ	12. Ⓐ Ⓑ Ⓒ Ⓓ	
13. Ⓐ Ⓑ Ⓒ Ⓓ		
14. Ⓐ Ⓑ Ⓒ Ⓓ		

Test Practice 4	Test Practice 5	Test Practice 6
1. Ⓐ Ⓑ Ⓒ Ⓓ	1. Ⓐ Ⓑ Ⓒ Ⓓ	1. Ⓐ Ⓑ Ⓒ Ⓓ
2. Ⓐ Ⓑ Ⓒ Ⓓ	2. Ⓐ Ⓑ Ⓒ Ⓓ	2. Ⓐ Ⓑ Ⓒ Ⓓ
3. Ⓐ Ⓑ Ⓒ Ⓓ	3. Ⓐ Ⓑ Ⓒ Ⓓ	3. Ⓐ Ⓑ Ⓒ Ⓓ
4. Ⓐ Ⓑ Ⓒ Ⓓ	4. Ⓐ Ⓑ Ⓒ Ⓓ	4. Ⓐ Ⓑ Ⓒ Ⓓ
5. Ⓐ Ⓑ Ⓒ Ⓓ	5. Ⓐ Ⓑ Ⓒ Ⓓ	5. Ⓐ Ⓑ Ⓒ Ⓓ
6. Ⓐ Ⓑ Ⓒ Ⓓ	6. Ⓐ Ⓑ Ⓒ Ⓓ	6. Ⓐ Ⓑ Ⓒ Ⓓ
7. Ⓐ Ⓑ Ⓒ Ⓓ	7. Ⓐ Ⓑ Ⓒ Ⓓ	7. Ⓐ Ⓑ Ⓒ Ⓓ
8. Ⓐ Ⓑ Ⓒ Ⓓ	8. Ⓐ Ⓑ Ⓒ Ⓓ	8. Ⓐ Ⓑ Ⓒ Ⓓ
9. Ⓐ Ⓑ Ⓒ Ⓓ	9. Ⓐ Ⓑ Ⓒ Ⓓ	9. Ⓐ Ⓑ Ⓒ Ⓓ
10. Ⓐ Ⓑ Ⓒ Ⓓ	10. Ⓐ Ⓑ Ⓒ Ⓓ	10. Ⓐ Ⓑ Ⓒ Ⓓ

Answer Key

Page 4
1. right
2. acute
3. acute
4. acute
5. obtuse
6. acute
7. acute
8. straight
9. acute
10. right
11. obtuse
12. obtuse

Page 5
1. acute
2. obtuse
3. acute
4. obtuse
5. acute
6. acute
7. right
8. acute
9. obtuse
10. straight
11. acute
12. obtuse

Page 6
1. 90° right
2. 30° acute
3. 180° straight
4. 120° obtuse
5. 20° acute
6. 40° acute
7. 150° obtuse
8. 40° acute
9. 170° obtuse
10. 60° acute
11. 35° acute
12. 180° straight

Page 7
1. 40° acute
2. 25° acute
3. 90° right
4. 130° obtuse
5. 180° straight
6. 20° acute
7. 30° acute
8. 45° acute
9. 70° acute
10. 10° acute
11. 110° obtuse
12. 90° right

Page 8
1. equilateral
2. right
3. scalene/obtuse
4. isosceles/acute
5. right
6. equilateral
7. isosceles/ obtuse
8. isosceles/ obtuse
9. scalene/acute

Page 9
1. 30°
2. 70°
3. 60°
4. 50°
5. 50°
6. 60°
7. 30°
8. 90°
9. 60°
10. 10°
11. 20°
12. 20°

Page 10
1. 30°
2. 40°
3. 40°
4. 40°
5. 75°
6. 70°
7. 60°
8. 65°
9. 55°
10. 20°
11. 40°
12. 30°

Page 11
1. square
2. hexagon
3. triangle
4. rhombus
5. rectangle
6. parallelogram
7. pentagon
8. trapezoid
9. octagon
10. parallelogram
11. isosceles triangle
12. trapezoid

Page 12
1. 90°
2. 120°
3. 90°
4. 110°
5. 110°
6. 60°
7. 125°
8. 80°
9. 100°
10. 130°
11. 100°
12. 75°

Page 13
1. 32 ft.
2. 48 cm
3. 40 m
4. 28 in.
5. 48 ft.
6. 52 yd
7. 44 cm
8. 84 mi.
9. 176 m
10. 308 mm
11. 564 ft.
12. 1,472 m

Page 14
1. 36 m
2. 42 ft.
3. 38 in.
4. 70 yd.
5. 56 ft.
6. 80 m
7. 70 cm
8. 256 mm
9. 234 ft.
10. 344 m
11. 444 yd.
12. 408 in.

Page 15
1. 40 ft.
2. 42 cm
3. 48 m
4. 64 in.
5. 74 m
6. 90 mm
7. 94 ft.
8. 76 cm

Page 16
1. 44 ft.
2. 51 m
3. 51 yd.
4. 58 cm
5. 28 cm
6. 230 in.
7. 220 in.
8. 153 ft.

Page 17
1. 60 in.
2. 60 cm
3. 32 m
4. 120 mm
5. 200 ft.
6. 300 m
7. 264 cm
8. 125 yd.
9. 750 ft.
10. 88 mi.

Page 18
1. 10 cm^2
2. 6 cm^2
3. 12 cm^2
4. 18 cm^2
5. 24 cm^2
6. 16 cm^2
7. 9 cm^2
8. 5 cm^2
9. 21 cm^2
10. 50 cm^2

Page 19
1. 20 sq. units
2. 27 sq. units
3. 16 sq. units
4. 33 sq. units
5. 10 sq. units
6. 18 sq. units
7. 26 sq. units
8. 24 sq. units
9. 36 sq. units
10. 15 sq. units

Page 20
1. 64 $in.^2$
2. 400 $ft.^2$
3. 225 m^2
4. 625 cm^2
5. 324 $yd.^2$
6. 121 cm^2
7. 3,844 $ft.^2$
8. 121 $miles^2$
9. 4.84 m^2
10. 1/4 $ft.^2$

Page 21
1. 70 $in.^2$
2. 72 cm^2
3. 180 m^2
4. 1,000 cm^2
5. 330 $yd.^2$
6. 700 mm^2
7. 3,150 $ft.^2$
8. 5,000 mm^2

Page 22
1. 320 $yd.^2$
2. 920 mm^2
3. 2,280 m^2
4. 5,580 $ft.^2$
5. 897 $in.^2$
6. 2,728 cm^2
7. 1,003 m^2
8. 6,232 mm^2

Page 23
1. 150 m^2
2. 1,500 cm^2
3. 900 $yd.^2$
4. 2,000 $ft.^2$
5. 540 mm^2
6. 2,600 $in.^2$
7. 3,600 $ft.^2$
8. 2,821 m^2

Page 24
1. 690 $yd.^2$
2. 2,108 $in.^2$
3. 1,550 m^2
4. 1,392 cm^2
5. 781 $mi.^2$
6. 704 $ft.^2$
7. 3,608 $in.^2$
8. 1,729 cm^2

Page 25
1. 48 $yd.^2$
2. 60 cm^2
3. 315 $ft.^2$
4. 72 cm^2
5. 595 m^2
6. 1,015 mm^2
7. 90 $yd.^2$
8. 1,720 cm^2
9. 137.5 $in.^2$
10. 156 $ft.^2$

Page 26
1. 60 $in.^2$
2. 440 cm^2
3. 2,460 $yd.^2$
4. 1,100 mm^2
5. 1,196 cm^2
6. 4,320 m^2
7. 288 $ft.^2$
8. 4,410 cm^2
9. 1,332.5 $in.^2$
10. 374 m^2

Page 27

1. 64 cm^3
2. 1,000 ft.^{3}
3. 1,728 yd.^{3}
4. 343 mm^3
5. 729 ft.^{3}
6. 512 cm^3
7. 8,000 mm^3
8. 15,625 yd.^{3}
9. 125,000 cm^3
10. 1/8 yd.^{3}

Page 28

1. 40 cm^3
2. 300 m^3
3. 210 in.^{3}
4. 144 cm^3
5. 384 m^3
6. 216 mm^3
7. 800 in.^{3}
8. 360 ft.^{3}

Page 29

1. 720 cm^3
2. 1,440 mm^3
3. 1,512 ft.^{3}
4. 1,440 yd.^{3}
5. 1,400 in.^{3}
6. 6,000 cm^3
7. 4,800 ft.^{3}
8. 2,160 mm^3

Page 30

5. r = 7 ft.
   d = 14 ft.
   C = 44 ft.
6. r = 18 cm
   d = 36 cm
   C = 113 cm
7. r = 11 cm
   d = 22 cm
   C = 69 cm
8. r = 21 in.
   d = 42 in.
   C = 132 in.

Page 31

1. C = 18.84 in.
2. C = 21.98 cm
3. C = 15.7 cm
4. C = 37.68 yd.
5. C = 43.96 m
6. C = 62.8 in.

Page 32

1. 50.24 in.^{2}
2. 113.04 m^2
3. 200.96 yd.^{2}
4. 314 cm^2
5. 153.86 ft.^{2}
6. 78.5 in.^{2}
7. 1,256 in.^{2}
8. 1,962.5 ft.^{2}

Page 33

1. 6.

3. 7.

11. 12.

Page 34

1. 7.

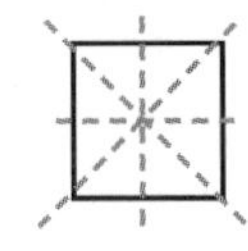

2. 8.

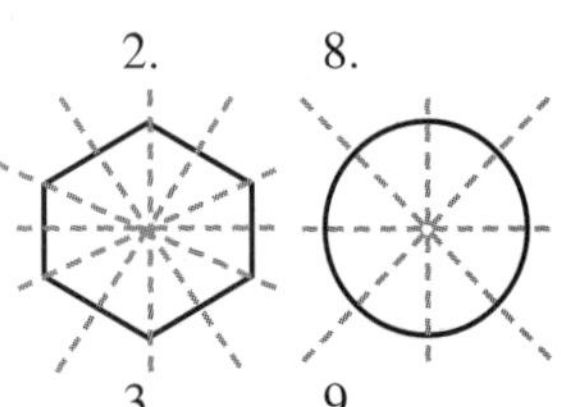

3. 9.

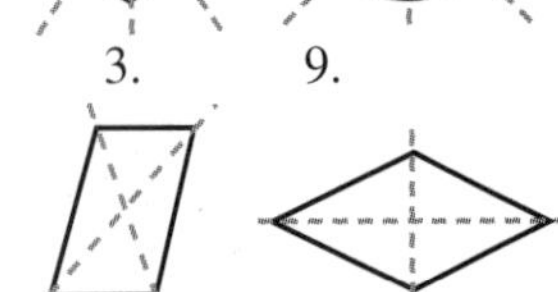

4. 10.

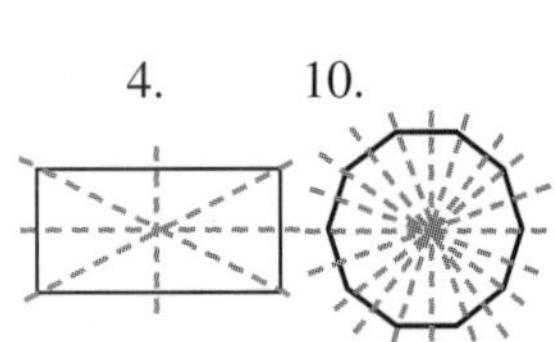

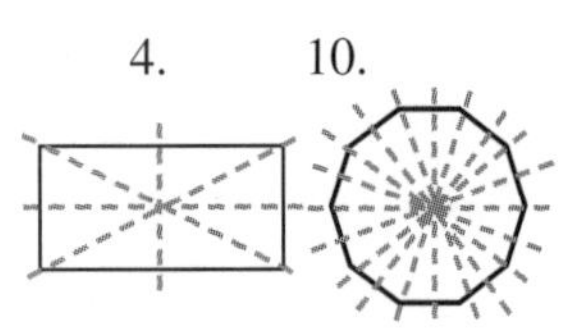

5. 11.

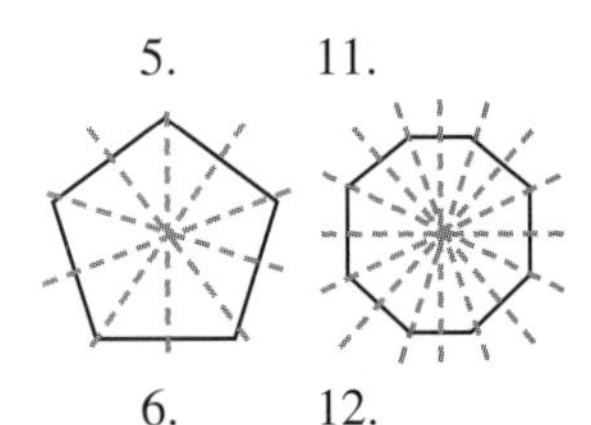

6. 12.

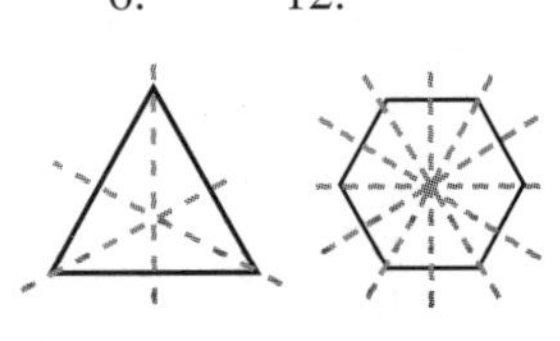

Page 35

1. AC
2. ABD
3. AB
4. AD
5. AB
6. AC

Page 35

1. AD
2. AB
3. AC
4. ACD
5. ABD
6. AB

Page 36

1. similar: A, B, C
   congruent: A, C
2. similar: A, C, D
   congruent: A, C
3. similar: A, B, D
   congruent: A, D
4. similar: A, B, D
   congruent: B, D
5. similar: C, D
   congruent: C, D
6. similar: A, B, D
   congruent: A, D

Page 38

1. triangular prism
2. rectangular prism
3. sphere
4. triangular
   pyramid
5. cylinder
6. square pyramid
7. cube
8. tetrahedron/
   triangular
   pyramid
9. cone
10. octahedron
11. icosahedron
12. dodecahedron

Page 39

1. square pyramid
   faces: 5
   edges: 8
   vertices: 5
2. triangular prism
   faces: 5
   edges: 9
   vertices: 6
3. cube
   faces: 6
   edges: 12
   vertices: 8
4. tetrahedron/
   triangular
   pyramid
   faces: 4
   edges: 6
   vertices: 4
5. rectangular prism
   faces: 6
   edges: 12
   vertices: 8
6. octahedron
   faces: 8
   edges: 12
   vertices: 6
7. dodecahedron
   faces: 12
   edges: 30
   vertices: 20
8. icosahedron
   faces: 20
   edges: 30
   vertices: 12

Page 40

1. A
2. B
3. D
4. C
5. B
6. C
7. B
8. B
9. A
10. A
11. B
12. A
13. C
14. A

Page 41

1. A
2. B
3. A
4. C
5. C
6. B
7. A
8. A
9. A
10. B
11. C
12. B

Page 42

1. A
2. D
3. C
4. B
5. B
6. A
7. C
8. A
9. C
10. C

Page 43

1. B
2. A
3. B
4. C
5. B
6. A
7. B
8. D
9. C
10. A

Page 44

1. C
2. B
3. A
4. C
5. B
6. A
7. B
8. A
9. A
10. B

Page 45

1. D
2. C
3. C
4. A
5. B
6. C
7. B
8. D
9. A
10. C